Teacher Education to Enhance Diversity in STEM

Addressing underlying issues in science education and teacher training, which contribute to continued underrepresentation of racial and ethnic minority students in STEM and STEAM subjects and careers, this timely volume illustrates how a critical postmodern science pedagogy (CPSP) can be used effectively to raise awareness of diversity issues amongst preservice teachers.

Using a case study design consisting of class observations, interviews, content analysis, questionnaires, and instructional interventions in preservice teacher training, the volume bridges science and multicultural education and investigates how curricular development and teacher preparation can be used to ensure that science education itself promotes diversity within STEM, and throughout education. Chapters also examine the intersections of science education and science literacy for both students and teachers and, in doing so, promote the importance of positive and accurate representation of diversity within science and research discourse. The book attempts to raise awareness regarding the need for meaningful curricular reform that creates real opportunities to address historical and scientific misinformation, while increasing diversity and inclusion in schools and society.

This important text will be of interest to postgraduate students, researchers, scholars, and preservice teachers in the fields of science and mathematics education, STEM, multicultural education, teacher education, urban education, and the sociology of education.

A. Anthony Ash II is a scholar, researcher, former science teacher, and a social justice activist. His research helps to inform inclusive practices surrounding science and multicultural perspectives.

Greg A. Wiggan is Professor of Urban Education, Adjunct Professor of Sociology, and Affiliate Faculty of Africana Studies at the University of North Carolina at Charlotte. His research addresses school processes that promote high achievement among urban and minority students, as well as science and issues of diversity. He has published over 100 publications, inclusive of 28 education books.

Marcia J. Watson-Vandiver is Assistant Professor of Elementary Education at Towson University. Her research interests explore various intersections of Black education, including resistance pedagogy, historical, and contemporary issues in urban education, multiculturalism, and transformative/emancipatory learning. She has published books and many articles on these topics.

Routledge Research in STEM Education

The *Routledge Research in STEM Education* series is home to cutting-edge, upper level scholarly studies and edited collections covering STEM Education.

Considering science, technology, engineering, and mathematics, texts address a broad range of topics including pedagogy, curriculum, policy, teacher education, and the promotion of diversity within STEM programmes.

Titles offer dynamic interventions into established subjects and innovative studies on emerging topics.

Books in this series include:

An Asset-Based Approach to Advancing Latina Students in STEM
Increasing Resilience, Participation, and Success
Edited by Elsa M. Gonzalez, Frank Fernandez, and Miranda Wilson

Teacher Education to Enhance Diversity in STEM
Applying a Critical Postmodern Science Pedagogy
Authored by A. Anthony Ash II, Greg A. Wiggan, Marcia J. Watson-Vandiver

For more information about this series, please visit: https://www.routledge.com/Routledge-Research-in-STEM-Education/book-series/RRSTEM

Teacher Education to Enhance Diversity in STEM

Applying a Critical Postmodern Science Pedagogy

A. Anthony Ash II
Greg A. Wiggan
Marcia J. Watson-Vandiver

Routledge
Taylor & Francis Group
NEW YORK AND LONDON

First published 2021
by Routledge
52 Vanderbilt Avenue, New York, NY 10017

and by Routledge
2 Park Square, Milton Park, Abingdon, Oxon, OX14 4RN

Routledge is an imprint of the Taylor & Francis Group, an informa business

Library of Congress Cataloging-in-Publication Data
Names: Ash, A. Anthony, II, author. | Wiggan, Greg A., 1976- author. | Watson-Vandiver, Marcia J., author.
Title: Teacher education to enhance diversity in STEM: applying a critical postmodern science pedagogy/A. Anthony Ash II, Greg A. Wiggan, Marcia J. Watson-Vandiver.
Description: New York, NY: Routledge, 2021. | Series: Routledge research in stem education | Includes bibliographical references and index.
Identifiers: LCCN 2020034671 (print) | LCCN 2020034672 (ebook) | ISBN 9780367621544 (hardback) | ISBN 9781003108184 (ebook)
Subjects: LCSH: Science teachers--Training of. | Science--Study and teaching--Social aspects. | Minorities in science
Classification: LCC Q181 .A746 2021 (print) | LCC Q181 (ebook) | DDC 507.1--dc23
LC record available at https://lccn.loc.gov/2020034671
LC ebook record available at https://lccn.loc.gov/2020034672

ISBN: 978-0-367-62154-4 (hbk)
ISBN: 978-1-003-10818-4 (ebk)

Typeset in Sabon
by KnowledgeWorks Global Ltd.

Contents

Tables

Figures

Abbreviations

CPSP	Critical Postmodern Science Pedagogy
ESL	English as a Second Language
MuEd	Multicultural Education
MuSE	Multicultural Science Education
NAEP	National Assessment for Educational Progress
NCES	National Center for Education Statistics
NSF	National Science Foundation
PCK	Pedagogical Content Knowledge
PISA	Program for International Student Assessment
PST	Preservice Teacher
REC	Race, Ethnicity, and Culture
TEP	Teacher Education Program
TIMSS	Trends in International Mathematics and Science Study
WMS	Western Modern Science
WSV	Wasetville (pseudonym)
UT	University of Tesaw (pseudonym)

1 The Critical Role of Science in Teaching Diversity

In the nearly eight decades following World War II, scientific innovation and population diversity have emerged as two of the most remarkable sociopolitical issues in the United States. While science has long been integral to human survival and progress, since the Soviet launch of Sputnik in 1957, it has come to permeate nearly every aspect of contemporary society (National Research Council [NRC], 2007a).[1] Increased applications of scientific knowledge have stimulated the creation of new industries, job growth, advances in healthcare, and methods to address environmental concerns (Tan, Barton, Gutiérrez, & Turner, 2012). Such advancements and prospects, however, belie the lack of U.S. workers trained to perform the highly skilled tasks associated with the rising number of science, technology, engineering, and mathematics (STEM) professions vital to sustaining this growth, as well as the nation's global economic stability and security (Gilbert, 2013; U.S. Bureau of Labor Statistics, 2015).

Recent employment projections predict that approximately 90% of the fastest growing occupations that require at least a bachelor's degree will rely most heavily on workers with extensive training in science and mathematics, and who can think critically to solve complex problems (Wang, 2013). However, because U.S. workers generally lack requisite knowledge in STEM disciplines that support the development of these skills, of the 123 million individuals needed to address this demand, only around 41% of the workforce will be adequately prepared to do so (Gordon, 2009; Wang, 2013). In fact, the *Brookings Institute* predicts that by the end of 2020, the demand for skilled technology jobs, for example, will outpace the supply of qualified applicants by one million (Duncan, 2017). As a result, U.S. businesses have increasingly requested H-1B visas to bring in STEM-skilled foreigners to fill this gap. Between the years 2001 and 2003 alone, the cap for the number of H-1B visas for highly skilled foreigners rose from 65,000 to 195,000,[2] widening the legal immigration channel by 40% (Chen & Yoo, 2010; Lowell & Martin, 2012). The inability to address the need for STEM professionals internally is but one dimension of the challenges faced by the United States in securing the future of its citizens.

Accompanying this rather unsteady progression into the 21st century is the racial and ethnic diversification of the U.S. population. In conjunction with the steady dissolution of institutional barriers throughout the 1950s and 1960s, the exponential presence of science in society helped to dramatically increase the number of cultural intersections, both in the United States and around the world. Facilitating this connection of people and ideas were developments in communication and transportation technology, which resulted in a surge of global migration, transnational dialogs, and interracial unions that increasingly blur the boundaries of race, ethnicity, language, and culture (Gilbert, 2013; Kellner, 2003, 2009; Lee & Bean 2010; Wang, 2012). For instance, recent census data predict that by 2060, the U.S. student demographic will reach 64% racial minority students, with biracial and multiracial students increasing at rates of 226% (U.S. Census Bureau, 2015). This raises questions about the role of science for discussing issues of diversity, as well as the potential impacts this could have on promoting positive and inclusive teacher practices (Humes, Jones, & Ramirez, 2011; Smelser, Wilson, & Mitchell, 2001), particularly in the understanding of race, language, and identity.

This rapid expansion of scientific innovation in the fields of medicine, information and communication technology,[3] and industry, along with the growing demographic heterogeneity is a strong indication that the global efficacy of the United States will necessarily rely on individuals from a variety of different racial, ethnic, and cultural backgrounds being trained in STEM fields. By extension, this places an enormous responsibility on the field of science education, which refers to "the scholarly and practical discipline concerned with the teaching, learning and assessment of science content, science processes and the nature of science" (McComas, 2013, p. 86). Though, despite being the fastest growing segment in the United States, racial and ethnic minority students, particularly those from urban areas, are least likely to pursue a STEM degree (Bianchini, 2012). In response to such issues, a large and growing body of science education research has emerged to raise STEM achievement and supporting the entry and persistence of underrepresented minorities in STEM areas (Allen & Eisenhart, 2017; Hurtado, Cabrera, Lin, Arellano, & Espinosa, 2008; Hurtado, Newman, Tran, & Chang, 2010; Riegle-Crumb & King, 2010; Riegle-Crumb, Moore, & Ramos-Wada, 2011). Reports issued by the National Academy of Sciences echo the critical nature of these trends and point to the need for substantial growth in the nation's workforce trained in STEM fields to support economic prosperity and high quality of life for this nation's citizens (NRC, 2007a, 2007b, 2010).

COVID-19

In December 2019, a novel coronavirus emerged in Wuhan, Hubei Province, China. Over the next several months, the virus spread around the world halting trade and travel, closing businesses and forcing layoffs, and leaving thousands sick and many dead (Centers for Disease Control and Prevention (CDC), 2020a). Though the purported origins of coronavirus disease 2019 (COVID-19 – *Severe acute respiratory syndrome coronavirus* 2) is either a wet market in Hunan, China, where live animals are sold for human consumption or in a laboratory, Chinese scientists at the Wuhan Institute of Virology dismissed these claims (Areddy, 2020). The subsequent transmission through global travel would prove to have devastating impacts on the global community. By March 11, 2020, the World Health Organization (WHO) had declared COVID-19 a pandemic (CDC, 2020a; WHO, 2020).

The scale and scope of this pandemic is a global crisis. By June 2020, the highly transmissible COVID-19 had infected over two million individuals in the United States and claimed the lives of over 100,000 (CDC, 2020b). Worldwide, these numbers reach nearly eight million infected and over 400,000 deaths in 216 countries (WHO, 2020). The lack of hospital beds and ventilators in intensive care units, as well as a shortage of personal protective equipment, also contributed to the massive increase in COVID-19 cases in just a few months (Jacobs, Richtel, & Baker, 2020).

While many vaccine trials are underway and a variety of prophylactics are under consideration, the disease continues relatively unabated and appears that without a vaccine it will be a source of contention for months, if not years, to come (Boulware et al., 2020; CDC, 2020a). Regardless of its impetus, populations the world over have been forced to find ways to contain and mitigate the transmission and effects of COVID-19 through social distancing (staying 2 m from another person) and wearing cloth or surgical masks to keep droplets containing the virus from spreading to others (CDC, 2020a).

Avoiding exposure, however, is less feasible for some. Those who work on the front lines during the pandemic (hospital workers, janitors, mail carriers, etc.) are at a greater risk of contracting and spreading COVID-19, a problem that is exacerbated in low-income and non-white communities (Reidmiller et al., 2018; Wallis, 2020). The latter is due in part to issues of poor health and access, such as getting to and from testing sites or getting a physician's order to have a viral test conducted (Tigue, 2020; Wallis, 2020). In Chicago, for example, while Blacks make up only 30% of the population, they represent more than 50% of the COVID-19 cases and roughly 70% of COVID-19 deaths (Yancy, 2020). Moreover, disproportionate numbers of racial minorities in jails, prisons, and detention centers point to a higher risk of contracting the disease (Wallace et al., 2020).

Today, science topics such as climate change, global warming, contraception, vaccinations, and environmental conservation yield political skepticism and division (New York State School Boards Association, 2017). Relatedly, with the recent onset of COVID-19, there is observable discord between political factions and scientific communities. In addition to the political divisions surrounding the perceptible threat and severity of COVID-19, racial tensions also heightened as Asian and Pacific Islanders have also been targets of discrimination, xenophobia, and intolerance (Huerta, 2020; Human Rights Watch, 2020). Here, both the federal government and civic agencies disseminated misinformation that discredited scientific research, and in many cases fueled vitriolic rhetoric (Yam, 2020).

Global Warming

In connection, another worldwide existential threat is global warming. Climatic fluctuations are common throughout the Earth's history; however, the past several decades have been exceptionally drastic. The past half-century has experienced the fastest rise in the average global temperature in recorded history, which have had marked effects on global climate patterns (MacMillan, 2016; NASA, 2020a; Nunez, 2019). The scientific consensus is that since the 1950s, human activity has been a large factor in the Earth's drastic climatic shifts (Cook et al., 2013, 2016; NASA, 2020a).

Issues

As fossil fuels are burned, carbon dioxide (CO_2) and other pollutants are released, heat is trapped in the atmosphere resulting in the melting of glaciers and ice sheets, warmer ocean temperatures, and less snow cover. The melting of snow and ice into the oceans has caused sea levels to rise over 8 inches since 1880, threatening coastal cities around the world (Church & White, 2011). The confluence of these events has also disrupted the thermal flows in the ocean, leading to an increase in dramatic weather events and environmental degradation (NASA, 2019, 2020b, 2020a).

Technological innovations such as weather satellites have contributed to our understanding of how and why anthropogenic influences related to fossil fuel production and consumption have contributed to the warming of the planet. As a result of increased fossil fuel use since the late 1800s, the average temperature of the Earth's surface has risen 1.62 degrees Fahrenheit, which intensifies snow and ice melting and increases ocean temperatures (NASA, 2019, 2020a). The two largest contributors of these so-called greenhouse gases are electricity production and transportation, both of which require substantial combustion of fossil fuels, the by-products of which are then released into the atmosphere (Environmental Protection Agency, 2015; MacMillan, 2016; NASA, 2020a).

Unabated, the trapped atmospheric heat and increased ocean temperatures and sea levels portend further natural disasters with increased frequency and intensity affecting millions of people. The nearly 50% of the U.S. population living in coastal regions, for example, will experience stronger, more intense hurricanes and storm surges (National Oceanic and Atmospheric Administration, 2018). And as the arctic and other polar regions lose ice, these areas will also be subject to a 1–8 ft seal level rise and increased flooding over the next 80 years. Some inland areas will experience prolonged drought, while more intense rainfall will trigger more landslides (NASA, 2019, 2020a).

Implications

These impacts are not only existential but carry steep economic consequences: from mitigation to public safety and healthcare responses, individual and company relocation, to cleanup and bioremediation (Smith, Schellnhuber, & Mirza, 2001; Stiglitz, 2006). Among the economic impacts are declines in the gross domestic product (GDP) of many countries due to social and environmental effects on agriculture, trade, forestry, fisheries, tourism, and recreation. While difficult to measure in some instances, the economic effects of global warming in the social sphere are not theoretical (Smith et al., 2001).

The longer the fossil fuel trend persists, the greater the (global) economic force required to combat the negative effects on societies and the environment in adapting to the forces of nature. And if history is any indication, these effects will be most acutely felt by low-income and minority communities, compounding pervasive inequalities they already face. This suggests a need for meaningful plans to address environmental inequities and strident efforts to enact them.

Environmental Justice

Also referred to as environmental racism, environmental inequality, or environmental injustice, environmental justice refers to a movement that began in Warren County, North Carolina in the 1980s that highlighted inequities in environmental exposures to pollution, subsequent effects on minority populations, as well as responses to these issues (Mohai, Pellow, & Roberts, 2009; Wenz, 1988). The focus here is the notions of justice and interconnectedness; that when the scarcity and desire for resources is such that some group must suffer, despite the negative impacts and the need for those most affected to help provide these resources, then there is no (environmental) justice (Wenz, 1988).

While initially aimed at addressing disparate levels of pollution exposure, such as the disproportionate number of toxic waste sites near minority communities, the focus of environmental justice now includes more expansive ways of understanding, interceding, and mitigating social inequities attached to the toxic environmental consequences of natural disasters, industrial pollution and radioactive contamination, and failing infrastructure on underserved communities (Chakraborty, Collins, & Grineski, 2016; Kyne & Bolin, 2016; Mohai et al., 2009).

In 2018, the National Climate Assessment (NCA) reported that climate change has and will continue to have the most demonstrable impact on marginalized and unserved groups. The report argues that due to historical inequalities related to infrastructure in marginalized and vulnerable communities, these groups will likely be most impacted by the overall disruptions related to health, safety, quality of life and economic growth caused by extreme weather events (Reidmiller et al., 2018).

The need for environmental justice is also evident in the primary inputs to these anthropogenic climate changes – fossil fuel production and consumption. In the industrial city of Detroit, Michigan, natural gas pollution induces asthma attacks and causes many missed days of school for thousands of Black children (Fleischman & Franklin, 2017).

When combined with other natural crises such as COVID-19, such threats against the health and public safety of minority communities are compounded. In Port Arthur, Texas, for example, nearly one-third of its citizens are African American are dually confronted with industrial pollution associated with the largest oil refinery in the country, as well as the disproportionate effects of COVID-19 (Tigue, 2020). The adverse effects of COVID-19 on the health and human safety of racial minorities are worsened by higher rates of respiratory illnesses, cancer, and cardiovascular diseases linked to industrial pollution (Correa, 2016; Reidmiller et al., 2018; Tigue, 2020). The cumulative effects of COVID-19, global warming, and environmental (in)justice suggest that further research and practices from a critical perspective are still needed.

Global Disruption or (Environmental) Justice for All

Critical postmodern science pedagogy (CPSP) is a useful lens through which to view the foregoing existential threats of COVID-19 and global warming, as well as approaches used to determine which populations are most affected by these threats and how to address related inequities, such as environmental justice.

With this critical, social-scientific perspective, unraveled are misconceptions and dismissals of minority communities, thus exposed is the inherent value and humanity of the individuals within them. It disentangles murky applications of science from the greed and consumption associated with the exploitation of minority communities and natural resources, and it interrogates assumptions about race as well as why and how racial minorities are predisposed to chronic disease, tend to be more incarcerated and are most impacted by environmental damage caused by a fossil fuel-forward economy.

Much of what is touted by commercial, industry-led innovations as progressive science presupposes infallibility; that it is good for the public insofar as it was developed and produced; that it has the final say in matters of public health and economic sustainability, and admonishes public debates and those in the scientific community who would seek to upend its preeminence. Such innovations often belie the premise of making the world a better place because the focus is not on understanding history or how the lack of inclusive research and development and lobbying for policies work for industry at the expense of the underserved.

A CPSP perspective exposes and decenters this narrative and any other that supports such commercial efforts to produce only that others might consume at a staggering rate and with no regard for the well-being of the global community. The ability of humans to subsist in the 21st century is directly proportionate to access to resources and the means with which to discern what is relevant and what is not regarding health and safety. This includes access to quality healthcare, affordable, nutritional food sources, and community infrastructure that addresses the current trajectory of global warming impacts.

From the tobacco industry to the fossil fuel industry, misleading forms of science ultimately offer little more than unwavering support for those who stand to accumulate wealth through (over)consumption of harmful products and exploitation of human capital and environmental resources, while the most vulnerable suffer. Unlike forms of science that seek to delude with one-off studies and solutions in search of problems, an authentic approach to science is open to and encourages interrogation of its methods and results.

This evokes Foucault's (1980) notion of the perennial struggle between truth and power, and the systematic process of ongoing self-interrogation of scientific knowledge and truth claims. Further, developing inclusive conversations and policies that reflect all communities provides access to positions within power structures to help facilitate necessary change. Though, lacking the ability to critique from within, economic convictions will continue to dominate social and environmental landscapes.

A CPSP perspective reveals that historical power structures upheld by the dominant class are evident in the all too frequent killings of Black men and women by law enforcement officers and subsequent racial unrest. Since 2014, the police-related deaths of Eric Garner, Michael Brown, Tamir Rice, Walter Scott, Alton Sterling, Philando Castile, Stephon Clark, Breonna Taylor, and George Floyd underscore the disregard and devaluation of Black bodies. CPSP suggests that this dehumanization does not happen in a vacuum but is facilitated by a void of a critical understanding of racial parity in society. Disrupting the paradigm of social stratification and subjugation is multifaceted and therefore requires deeper, more holistic analyses to cultivate a truly inclusive society.

In this way, a critical view of science and society highlights an interminable link between environmental injustice and social injustice. Both social and environmental factors are at play when underserved communities distrust law enforcement and are forced to bear the brunt of unfettered industrial and technological "progress" and pollution. In this sense, sustainability is not only a desired outcome of environmental justice but a necessary one (Hornik, Cutts, & Greenlee, 2016).

Fossil Fuels versus Renewable Energy

The insistence on fossil fuel production and consumption is intimately involved with state and personal wealth accumulation and an exploitation of natural resource scarcity. Since the early 1800s, fossil fuels such as coal, crude oil, and natural gas have been an anchor for the development of the United States into an industrialized country (Höök & Tang, 2013; Jones, 2010). Given the exponential growth of industrial societies around the world and that no country has every natural resource it needs, global interdependence on such resources would become a significant aspect of 19th-century global economy (Sanders, Sandvik, & Storli, 2019).

Even now, oil and natural gas comprise more than 66% of U.S. energy needs (Ladislaw & Barnett, 2019). However, the push for and reliance upon crude oil extraction and production has underlain many geopolitical struggles, namely, involving the Organization of the Petroleum Exporting Countries (OPEC). Since its formation in 1960, the OPEC has been an important player in the fossil fuel industry with enormous market power. With the ability to set crude oil prices and production levels, OPEC has the influence to impact global economies through embargoes, oil shocks, and recessions (Brew, 2019; Hamilton, 1983).

Impacts

However, increasing data and research since the 1980s have demonstrated the negative effects of fossil fuel procurement and consumption on the environment, with direct links to human consumption and waste. Moreover, human population growth trends suggest that fossil fuel scarcity will become an important issue in future decades. As countries around the world increasingly face extreme weather events linked to anthropogenic climate change, debates regarding the deleterious effects of fossil fuels on the environment and health of global populations have intensified (Höök & Tang, 2013).

Recent innovations in renewable energy have sought to offset these impacts. Building on the past several decades of research, 21st-century technological innovations have opened paths toward a more sustainable future based on the use of renewable energy. Renewable forms of energy in the form of solar, wind, and water generating technology offer an alternative to this self-destructive trajectory (Nesta, Vona, & Nicolli, 2014). By focusing on innovations that provide renewable methods of energy production, many of the burdens posed on humans, flora, and fauna can be attenuated.

Incentives

Another benefit of using renewable energy is economic in nature. In the 1950s, U.S. taxpayers helped subsidize the petroleum industry (Brew, 2019). Currently, the U.S. government provides tax incentives and grants for companies that produce renewable energy infrastructure and tax credits for end users of such technology (Cunningham, 2019; U.S. Energy Information Administration, 2019). However, Owen (2004) argues that to do this, fossil fuel subsidies would need to be removed so that the pricing more accurately reflects their market value relative to the environmental damage they cause.

A potential approach might include using biofuels derived from plant and animal material as a market disruptor to help offset OPEC power, which could slow global production and consumption of crude oil-based fuel and lead to more sustainable fuel choices (Hochman, Rajagopal, & Zilberman, 2010). And yet, the efficacy of such an approach is contingent on changing public perceptions about fossil fuels and the reliance thereon (Wüstenhagen, Wolsink, & Bürer, 2007).

This suggests a critical, holistic approach that includes urging science and industries to bolster innovation and production of renewable energy sources, informing the public, offering precisely targeted incentives, and creating barriers to the extraction, manufacture, and consumption of fossil fuels can help move the global conversation and action toward the necessary implementation of renewable energy. One such barrier is the use of a carbon or environmental tax (Metcalf, 2020; Stiglitz, 2006). Economists use computer models that examine links between the economy and climate to assess the damaging impacts of climate change. These integrated assessment models (IAMs) are used to calculate a carbon tax rate that would help curtail the types of industrial and economic activities that leave lasting damage to the environment and global economy (Metcalf, 2020). Many economists agree that a carbon or environmental tax is one of the best ways to stem the use of fossil fuels and redirect efforts toward more sustainable forms of energy (Metcalf, 2020; Stiglitz, 2006).

The lack of positive and negative incentives for reducing the reliance on fossil fuels along with the already disproportionate effects of environmental degradation on racial minorities suggests ongoing disparities in healthcare.

Racial and Class-Based Inequities in Healthcare

Ecological effects on the human population due to changes in climate, natural disasters, and pandemics highlight the need for a durable healthcare system. However, given the deep-rooted and divisive nature of racism in this country, non-whites are often overrepresented in terms of medical issues, though undercounted in terms of providing necessary public health resources (Agin, 2006; Wallis, 2020). The lasting outcomes of discrimination and oppressive attitudes toward minority communities illustrate how issues of healthcare access are clearly delineated along the lines of race and class.

Although contemporary societies strive to provide equal access to substantive, quality, and affordable healthcare, research shows that income, race and ethnicity, and social class can determine the extent to which this is a reality for minority groups (Essink-Bot, Lamkaddem, Jellema, Nielsen, & Stronks, 2013; Kaplan, Pamuk, Lynch, Cohen, & Balfour, 1996; Kennedy, Kawachi, & Prothrow-Stith, 1996; Nelson, 2002). And, generally speaking, those with higher incomes tend to have better health (Agin, 2006; Kawachi & Kennedy, 1999; Wilkinson, 1996). However, racial minorities have higher death rates from cardiovascular disease, diabetes, and cancer, regardless of type of insurance, income, age, or the severity of the medical condition (Nelson, 2002; Wallis, 2020). And yet, there appears to be a systemic apprehension toward providing the type of healthcare necessary to support a high quality of life (Nelson, 2002).

Race

Race is indeed a factor in providing public health services. Studies have shown, for example, that appropriate pain diagnoses and management in cardiac catheterization are skewed toward white men at the expense of black women in particular, and interestingly, irrespective of the race of the physician (Bertoni et al., 2005; Burgess, Van Ryn, Crowley-Matoka, & Malat, 2006; Chen, Rathore, Radford, Wang, & Krumholz, 2001; Schulman et al., 1999). One implication is that physicians' attitudes and biases influenced whether certain individuals received what is often life-saving care, thus inherently segmenting the population by race regarding healthcare.

COVID-19 disproportionately affects non-whites, who tend to work jobs that put them more at risk (Wallis, 2020). And despite an overall decline in incarceration rates in U.S. prisons, non-whites remain overrepresented in the country's nearly 5,000 correctional and detention facilities, and are therefore more likely to contract the disease in these crowded environments, which often lack adequate medical resources (Gramlich, 2020; Wallace et al., 2020; Wallis, 2020). In addition to increased exposure to and lack of protection against the disease, blacks are more than twice as likely to die from the disease than whites because of a propensity toward chronic diseases such as heart disease, hypertension, and diabetes, which are linked to limited healthcare access and healthy food options (Agin, 2006; Nelson, 2002; Wallis, 2020).

Class

Social class is another determinant of receiving high-quality, affordable healthcare. Poverty is not only a barrier to accessing adequate healthcare but is also an influential factor in the frequency and extent of illness (Agin, 2006; Wallis, 2020; Wilkinson, 1996, 2007). As a result, from rural to urban America, illness and disability are far more prevalent and life expectancy is shorter for lower income individuals (Agin, 2006; Wilkinson, 1996). Rather than leveraging the full weight of public health as applied science to support a healthy society, healthcare exists as part of an imbalanced and politically driven free-market system that stymies high quality of life for many low-income individuals (Agin, 2006). And because non-white individuals tend to earn less than their white counterparts, those already beleaguered by social stresses, mistrust of institutions, and inadequate sources of healthy foods are also most susceptible to healthcare inequities (Wallis, 2020; Wilkinson, 2007).

Growing environmental challenges require a greater need for healthcare systems that serve all people, but particularly those in marginalized communities who are most often beset with issues of access and affordability. Along with the disproportionate effects of environmental (in)justice, this suggests that receiving high-quality healthcare to reverse this course is an important and immediate concern for people in marginalized communities.

The vitality and posterity of the United States necessitate larger, more critical STEM discussions (Priest, 2014; Roth & Lee, 2004). As such, schools serve as epicenters for positively reshaping and informing citizenry. In the ostensible pursuit of national security, economic sustainability, and high quality of life,[4] the combined influences of science and diversity hold profound impacts for public education in the 21st century.[5]

Science Achievement and Teacher Preparation

As schools and society grow more diverse, the onus for preparing a democratic and critically thinking workforce falls largely on the nation's future educators. In the interest of growing a diverse and competitive workforce for global economic sustainability, educational research has subsequently focused on teacher preparation (NRC, 2007a, 2007b, 2010;

Riegle-Crumb & King, 2010). Due to this presumed link between education and the economy, teacher education programs (TEPs) have come under increased pressure to produce teachers who can engage students in diverse classrooms and prepare a highly skilled workforce capable of addressing the nation's economic needs (Cochran-Smith, 2010; Cochran-Smith & Zeichner, 2009; Darling-Hammond, 2010, World Economic Forum, 2020).

To some degree, this perceived threat to the nation's social, economic, and political efficacy in the global context is also due to the large investments in STEM education and research made by other nations (Bianchini, 2012). Literature on students' decisions to pursue a STEM degree indicates that issues of access, equity, and relevance continue to have a crucial impact on their chances to secure employment in these areas (Chen, 2009; Correll, 2001; Nembhard, 2005; Riegle-Crumb & King, 2010; Xie & Shauman, 2003). This research has shown, for example, that students attending high schools in low-income urban areas tend to be ethnoracial minorities who do not feel academically prepared in science as they transition into postsecondary settings (Lee, Robinson, & Sebastian, 2012; Toldson & Lewis, 2012). Moreover, these schools generally lack the resources needed for high student achievement, such as qualified teachers, classroom and laboratory materials for science competency, and culturally responsive instruction (Darling-Hammond, 2005; Gay, 2002; Johnson, Kahle, & Fargo, 2007; Kozol, 2005; Ladson-Billings, 2009; Moscovici, 2008). Given this privation of resources, quality instruction, and opportunities that promote student achievement, the notion of a knowledge-based economy becomes especially troubling for diverse schools, and particularly for those in low-income urban communities (Darling-Hammond, 2005; Nembhard, 2005).

It is important to note, however, that low student achievement in racially diverse and/or underserved communities alone cannot explain the nation's overall middling STEM education programs and the underrepresentation of minorities in these fields. Nor does it explain the relatively underwhelming performance of American students on international assessments such as the *Trends in International Mathematics and Science Study* (TIMSS) and *Program in International Student Assessment* (PISA) (Darling-Hammond, 2010; Organisation for Economic Co-operation and Development [OECD], 2011a, 2011b, 2013a, 2014a; PISA, 2018a, 2018b).

The latest PISA (2018a, 2018b) assessment compared 15-year-olds' achievement from various *Organisation for Economic Co-Operation and Development* (OECD) nations in the fields of mathematics, science, and literacy. Related to this study, the United States ranked 37th in science and 18th in mathematics according to the latest PISA (2018a, 2018b) assessment. It is important to note that students in the United States performed above the OECD average in reading and science, and below the OECD average in mathematics. According to the latest PISA (2018a, 2018b), U.S. trends in reading since 2000, mathematics since 2003, and science since 2006 are "stable, with no significant improvement or decline" (p. 1). Socioeconomic predictors aligned with other OECD averages, demonstrating that socio-economically advantaged students typically outperformed disadvantaged students in reading, mathematics, and science. However, even with controlling for socioeconomics, access to quality teaching was an indicator of student success. The PISA 2018 report suggests:

> In the United States, 48% of students enrolled in disadvantaged schools and 5% of students enrolled in advantaged schools attend a school whose principal reported that the capacity of the school to provide instruction is hindered at least to some extent by a lack of teaching staff. On average across OECD countries, 34% of students in disadvantaged schools and 18% of students in advantaged schools attend such a school. (2018, p. 7)

Classroom behavior was another indicator of student performance. Students in disruptive classrooms scored thirty-three (33) score points lower in reading than their counterparts, even after accounting for socio-economic status.

Where, even after controlling for socioeconomic backgrounds, as measured by the PISA index of social, cultural, and economic status, U.S. students' science scores showed very little improvement (OECD, 2014a, p. 230). While analyses of 2000, 2009, and 2018 PISA data show that controlling for SES did result in mathematics and reading gains among "socially advantaged U.S. students," that there were increases of any significance simultaneously highlights issues of educational access and opportunity in this country (Carnoy & Rothstein, 2013; Rebell & Wolff, 2008). This, although over the last decade, the United States has been among the top five OECD countries in annual per-pupil expenditures (NCES, 2019a; OECD, 2013b, 2014b). Comparative assessments administered by PISA to OECD countries and districts demonstrate that many nations are outperforming American students in science, math, and reading; three subjects crucial to education and social mobility in a knowledge-based global economy (Cromley, 2009; OECD, 2011a, 2013a). While the recent 2018 projections show stabilization in each of these subject areas, out of 65 countries,[6] in math, U.S. students dropped from 31st to 37th in math and increased from 24th to 17th in science (OECD, 2013a, 2014a; PISA, 2018a, 2018b).

In addition to the latest PISA (2018a, 2018b) study, the Trends in International Mathematics and Science (TIMSS) (2015a, 2015b) demonstrates fourth- and eighth-grade students' performance. As displayed

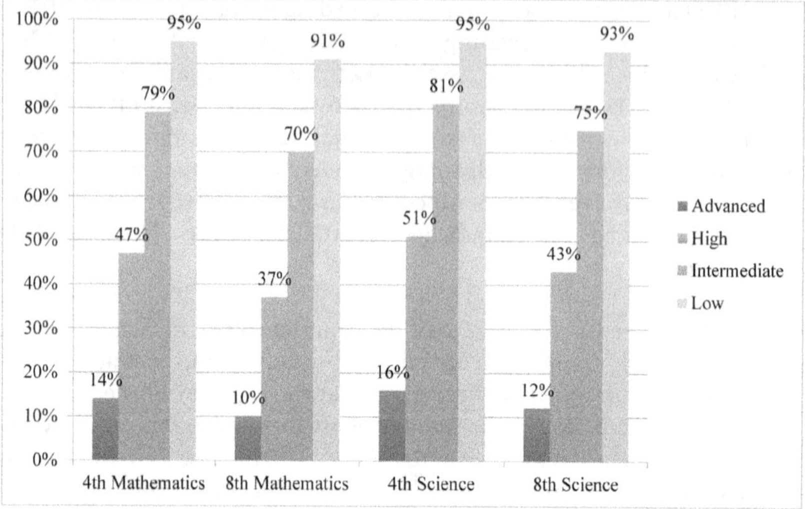

Figure 1.1 Percentage of U.S. fourth- and eighth-grade students reaching the TIMSS international benchmarks in science and mathematics: 2015

Source: Trends in International Mathematics and Science Study (TIMSS) (2015a, 2015b).

in Figure 1.1, fourth- and eighth-grade student success shows progressively lower numbers of students showing advanced and high proficiencies in mathematics and science. It is important to note that Figure 1.1 displays progressive percentages of students reaching various TIMSS benchmarks. While over 70% of U.S. fourth- and eighth-grade students demonstrate an intermediate competency across mathematics and science, advanced level proficiency is much more limited.

The above data report U.S. achievement in comparison to other participating nations and international districts. Findings from these international assessments are helpful when putting U.S. student achievement in mathematics and science into a global perspective. Notwithstanding, it is important to mention that international assessments have received some warranted criticism in educational research (Torney-Purta & Amadeo, 2013). Researchers note the unavoidable inconsistencies when comparing international educational systems (ibid.). The *Program for International Student Assessment* (PISA) (2018a) notes that in the United States, "low- and high-performing students are clustered in the same schools less often than the OECD average" (p. 6). Additionally, researchers also problematize student selection processes in countries where educational access is not available to all citizens (Singer & Braun, 2018). To that end, it is also important to examine student achievement on a national level. The *National Assessment of Educational Progress*

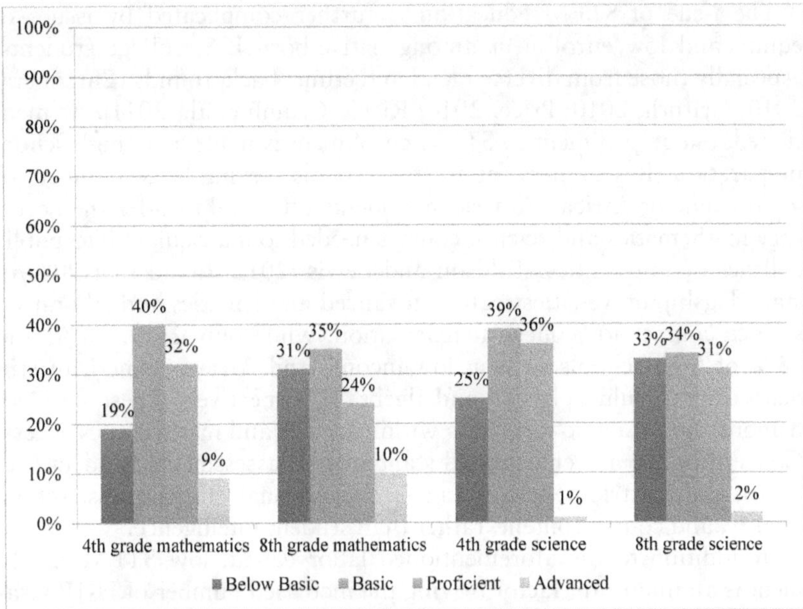

Figure 1.2 National fourth- and eighth-grade NAEP achievement-level results by percentage proficiency in mathematics and science: 2015 (science), 2019 (mathematics)

Source: NAEP (2015a, 2015b, 2019a, 2019b).

(NAEP) presents the latest fourth- and eighth-grade student achievement data. This report measured the percentage of students across proficiency levels in mathematics and science (2015a, 2015b, 2019a, 2019b), as displayed in Figure 1.2.

As demonstrated in Figure 1.2, U.S. student performance on standardized mathematics and science assessments is concerning. In mathematics, 41% and 34% of students are at or above proficiency, respectively. These numbers are comparable to science performance, which reports 37% and 33% proficiency in fourth and eighth grades, respectively. However, it is important to note that science results are relatively dated, as the latest reported NAEP trends are from 2015. To that end, additional reports also provide details on U.S. student performance. According to the National Center for Education Statistics' (NCES) (2018) *Condition of Education*, 40% of U.S. twelfth graders scored below basic proficiency in science, while only 60% demonstrated basic proficiency and 22% were above proficient. With high school preparation being a direct indicator for the STEM labor workforce (Israel, 2017), these statistics provide a glaring look into the status of science education and advancement in the United States.

The issue of STEM education is further complicated by issues of equity and low enrollment among native-born U.S. college students, especially those from diverse racial and ethnic backgrounds (Ehrenberg, 2010; Griffith, 2010; Price, 2010; Riegle-Crumb et al., 2011). As mentioned, one impediment to STEM enrollment is inadequate high school preparation. Researchers found that schools serving large numbers of low-income or African American students often fail to offer the necessary mathematics and science courses needed to matriculate into public colleges or universities. Toldson and Lewis (2012) found that 90% of state flagship universities require advanced mathematics and laboratory science courses for college matriculation, while only 29%, 40%, and 65% of high schools serving low-income and African American students offer calculus, physics, and algebra II, respectively. These statistics demonstrate systemic inequities within science and mathematics access. Concerning the aforementioned standardized assessments, student success is more attributed to teacher's instructional effectiveness, school quality, and course content, rather than student intelligence.

In addition to the aforementioned labor deficit, low STEM enrollment is an important factor driving the increased number of H-1B visas requested by U.S. companies for foreign professionals trained in these areas (Gordon, 2009; Lowell & Martin, 2012; Wang, 2013). Varying levels of student preparation and prior educational experiences help to explain differences in their rates of persistence in attending colleges and universities (Griffith, 2010). In addition, enrollment in Advanced Placement Biology and Advanced Placement Chemistry significantly influences students' choice of attending a postsecondary institution and helps prepare them for the challenges of college science courses (Hoepner, 2010). Evidence also shows that high school students who enroll in more advanced science courses are more likely to feel prepared in science regardless of their race or ethnicity (Israel, 2017). However, recent data suggest that only around 18% of U.S. undergraduates receive science and engineering degrees (NCES, 2019b), in comparison to China's 40% (McCarthy, 2017). China and India currently operate as the frontrunners of college graduates, totaling over 7.3 million STEM graduates combined (ibid.). Research estimates predict that by 2030, Chinese graduates aged 25 to 34 are expected to rise 300% in comparison to 30% in the United States (ibid.). Since the beginning of the 21st century, local and international metrics such as these underscore the reality that the United States consistently lags behind many so-called developed countries in terms of quantity and quality of K–12 STEM education (Darling-Hammond, 2010; Ong, Wright, & Orfield, 2011; World Economic Forum, 2020).

The National Science Foundation's (2020a) *State of U.S. Science and Engineering* captured recent statistics concerning the country's international performance across several indicators. Figure 1.3 demonstrates an

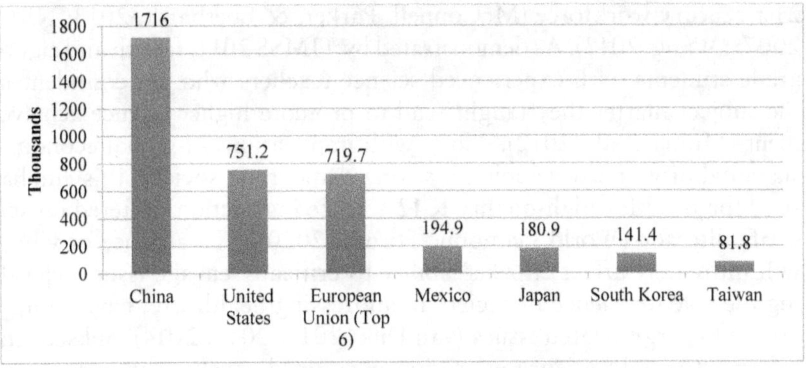

Figure 1.3 First university degrees in science and engineering, by selected region, country, or economy: 2015

Source: National Science Foundation (2020a).

international comparison of top countries regarding postsecondary science and engineering degrees conferred.

As noted in Figure 1.3, China largely leads in international science and engineering degrees conferred, with the United States and the top six European Union countries collectively following. The remaining countries produced between 80,000 and 195,000 science and engineering degrees in 2015: Mexico, Japan, South Korea, and Taiwan (National Science Foundation, 2020a).

While nations such as China, South Korea, and France continue making rapid advances in science and technology-related fields, U.S. children's interest in and academic preparation for STEM careers appears to be declining (McCarthy, 2017). These results support TIMSS 2015 findings from student assessments in 59 countries, which indicate a strong, bidirectional influence of students' attitudes toward science and science achievement (Mullis, Martin, Foy, & Stanco, 2012). These data show that positive attitude about science was a strong indicator for high science achievement, with the converse also being true (Mullis et al., 2012). With respect to science achievement, PISA findings further suggest that the poor performance could be attributed to the fact that the number of students with positive attitudes about science (i.e., factors such as interest, anxiety, perceived value, self-esteem, and self-efficacy) has also declined (Mullis et al., 2012; PISA, 2015).

Among other such indicators, TIMSS and PISA have prompted questions about the nation's comparative advantage in STEM, as well as the overall integrity of its educational system, particularly in the preparation of high-quality teachers (Darling-Hammond, 2005, 2010; World Economic Forum, 2020). The consensus in science education scholarship is that thorough content knowledge is vital to preparing a highly skilled,

21st-century workforce (McConnell, Parker, & Eberhardt, 2013; NRC, 2007a; Wang, 2013). As demonstrated by TIMSS 2011, fourth and eighth grade students with experienced science teachers who are confident in the subject matter they taught tend to promote higher science achievement (Mullis et al., 2012). Along with implications of global economic sustainability and developing a strong democratic society, this emphasizes the need for high-quality K-12 science instruction centered on scientific literacy (World Economic Forum, 2020). The absence of which will ultimately affect citizens' ability to critically engage issues regarding the role of science in society in addressing healthcare, environmental, and energy-related issues (van Dijk, 2011, 2013, 2014). Subsequent reform efforts in science education positioned scientific inquiry as an essential component of science instruction for greater scientific literacy and proficiency (Park, Jang, Chen, & Jung, 2011).

Students involved in inquiry-based learning seek to understand scientific ideas as they occur in nature, as well as the methods by which scientists study these phenomena (Stone, 2020). Science educators who teach from an inquiry perspective seek to create learning opportunities for students to develop evidence-based scientific questions and explanations (Gilbert, 2013; Stone, 2020). In this sense, effective science teaching that consistently, if incrementally, moves toward science proficiency involves an active process in which students explore natural phenomena by generating questions and theories related to their personal experiences, as well as examine ideas generated within the larger scientific community (Park et al., 2011). To this end, the *National Research Council* (NRC) offers a framework K-8 science proficiency and curriculum design to help support classroom teachers in developing students' scientific literacy. The NRC's framework, since updated in 2012, comprised four fundamental elements of learning, states that students who understand science:

1 Know, use, and interpret scientific explanations of the natural world.
2 Generate and evaluate scientific evidence and explanations.
3 Understand the nature and development of scientific knowledge.
4 Participate productively in scientific practices and discourse (NRC, 2007c, p. 334).

The four learning strands encompassed by this framework are positioned as learning goals that address the types of scientific knowledge, reasoning skills, and practices essential for the development of students' scientific literacy, and thus their ability to demonstrate science proficiency. The development of science proficiency is most effective when the focus of curriculum and instructional design is on the interplay of these four strands, which requires strong content knowledge as well as the ability to deliver effective instruction (NRC, 2012). Moreover, the demonstrable ability of scientific innovation to address societal needs suggests that scientific knowledge will continue to be a significant aspect of education in years to come. Yet, due to the lack of content knowledge, many science teachers are unable to support high levels of achievement among students who might otherwise pursue careers in science-related fields and contribute to in this capacity (Buczynski & Hansen, 2010; McConnell et al., 2013).

Johnson and Marx (2009) similarly argue that "science teacher quality is a concern because many science teachers feel that they were not prepared with adequate content knowledge or the instructional skills to teach science effectively" (p. 117). On the other hand, Hume and Berry (2011) report that science graduates may also contribute to poor teaching quality because they underestimate the cognitive demands and pedagogical reasoning needed to support student achievement. Thus, it is unlikely that students will become proficient in science simply by developing pedagogical strategies or by subject matter content alone, which suggests that effective science teaching is a multidimensional endeavor that requires ongoing development (McComas, 2014; NRC, 2007c; Schneider & Plasman, 2011).

The growing complexity of issues related to science, society, and diversity has stimulated a considerable amount of research interested in culturally responsive teacher preparation, as well as science teaching and learning (Emdin, 2016; Gilbert, 2013). And because science is still dominated by a male Eurocentric perspective, this literature has increasingly sought to address the need for a diverse and highly skilled workforce through conceptual and pedagogical methods such as critical pedagogy, culturally responsive pedagogy (CRP), as well as multicultural science education (MuSE) (Atwater, 1989, 1996a, 2011; Barton & McLaren, 2001; Basu & Barton, 2010; Hodson, 1993, 1999; Hodson & Dennick, 1994; National Association for Multicultural Education, 2019). For example, the National Science Foundation's (2020b) *State of U.S. Science and Engineering* reported the number of underrepresented minorities in science and engineering fields relative to the general U.S. residential population. Figure 1.4 provides a clearer depiction of these trends.

While underrepresented minorities constitute 28.1% of the U.S. adult population, they represent only 13.3% of the science and engineering occupations. Similarly, Blacks and Hispanics constitute 11.9% and 15.6% of the total population, respectively, while they represent only 5.6% and 7.5% of science and engineering occupations. This underrepresentation should be explored. We posit that science education is the stimulus for promoting more diverse representation in STEM fields.

Thus, in addition to exposing students from all backgrounds to STEM at an early age, this research suggests that using critical and culturally responsive approaches such as hip-hop to engage students, as well as

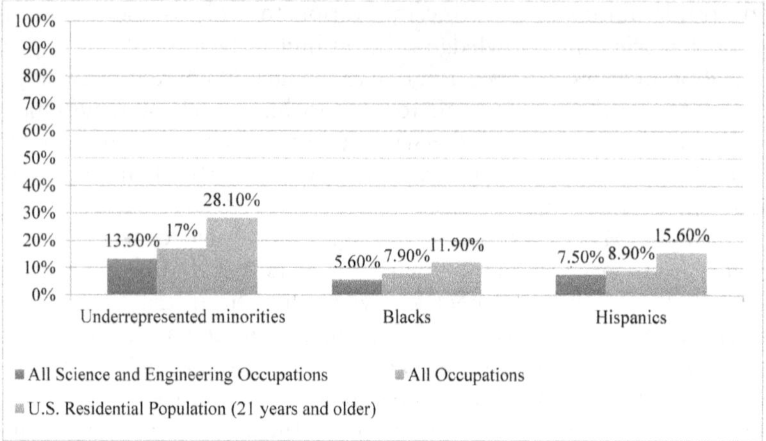

Figure 1.4 Underrepresented minorities, Blacks, and Hispanics in science and engineering in comparison to all occupations: 2017

Source: National Science Foundation (2020b).

place-based methods that integrate science into the community can help promote science achievement and thus mitigate the underrepresentation of minorities in these areas (Emdin, 2016). While they tend to be domain-specific, these and other culturally inclusive curriculum enhancements and pedagogical methods are important aspects of supporting teacher preparation for minority student achievement in science as well as other content areas (Banks & Banks, 2004, 2010; Gay, 2002; Ladson-Billings, 2006). What is not yet clear, however, is the potential impact of science as a transdisciplinary approach to initiate critical discussions about racial and ethnic diversity.

In general, the number and the success of TEPs that effectively prepare preservice teachers (PSTs) to teach in diverse settings is also unclear (Kumar & Hamer, 2012; Sleeter, 2001; Zeichner, 2003). Likewise, two decades of research on multicultural teacher education have at best produced ambiguous results. While some multicultural teacher education courses and related field experiences have positively impacted preservice teachers' dispositions others have shown little to no influence (Bennett, 2013; Delany-Barmann & Minner, 1997; Garmon, 2005; Ross & Smith, 1992; Sleeter, 2001). A majority of research suggests that PSTs are at the very least, underprepared for the diverse settings that many of them will enter upon graduation (Cho & DeCastro-Ambrosetti, 2005; Howard, 2006; Swartz, 2003). Among many possible explanations for the lack of progress, research suggests that this stems from: White PSTs entering TEPs with little cross-cultural and intercultural knowledge; infrequent or constrained discussions about race and ethnicity; and overly narrow applications of multicultural and culturally responsive approaches that tend to elide more critical and comprehensive analyses of human history (Carter, 2004; Cochran-Smith, 1995; Gilbert, 2013; Hickling-Hudson, 2004; Sleeter, 2001; Swartz, 2005). Notwithstanding, the *National Association for Multicultural Education* contends the intersections of scientific literacy and diversity are inseparable. In fact, many multicultural educators suggest that science education traverses into matters of classroom justice (National Association for Multicultural Education, 2019), considering that many presuppositions about science's distancing from issues of culture and diversity are false, especially in practical classroom settings (Durodoye, 2003; Emdin, 2010).

Problem Statement

The central problem is that TEPs do not explicitly address many key issues needed for PSTs to develop deeper, more critical understandings of racial and ethnic diversity that support student agency and achievement in diverse settings. Student populations are increasingly diverse and yet the teaching force remains largely comprised of "[w]hite, monolingual, middle class, and female" whose lived experiences tend to be appreciably different from the student populations they serve (Milner, 2006, p. 344). Nearly 80.1% of the 3.85 million schoolteachers in the United States are White (NCES, 2017), 76% of whom are female, whereas the growing majority of students they teach are from various ethnoracial (ethnic/racial) backgrounds (National Center for Education Statistics [NCES], 2017). Research has consistently shown that many in this cohort often find it difficult to make the necessary connections to reach students with backgrounds different from their own (Delpit, 2006; Emdin, 2016). Whether the result of social or cultural misconceptions about race, many neophyte educators are not prepared to teach in diverse settings (Cho & DeCastro-Ambrosetti, 2005; Ladson-Billings, 2000; Swartz, 2003). However, as they complete their education programs, many PSTs will likely find jobs at diverse schools, where they will be surrounded by socioeconomic and cultural characteristics with which they are unfamiliar (Swartz, 2003). Thus, despite the shift toward no single racial majority, many PSTs will be ill-equipped to meet the needs of an evolving and diverse U.S. society (Howard, 2006). Notwithstanding, due to the homogeneous nature of STEM disciplines, this shift toward diversity and multicultural perspective is increasingly important for science teachers to embrace (Avraamidou, 2018; Le & Matias, 2018).

Implicit within these phenomena, as well as the relative silence about issues of race and ethnicity in TEPs, is that the inadequate preparation of PSTs ultimately results in negative outcomes for non-White students. Multiracial and multicultural classrooms are increasingly led by teachers who find it difficult to reach their students, a costly disconnection that often results in poor instruction and low academic achievement (Brown, 2007; Delpit, 2006; Howard, 2006). The inability to connect with students, in turn, perpetuates educational inequality through what King (1991) refers to as "dysconscious racism," or the implicit sense of privilege and social dominance that engenders deficit-based thinking about the intellectual capacity of non-White students (Brown, 2007; Irvine & York, 1993; Milner, 2006; Sheth, 2018). This suggests that understanding issues of race and ethnicity from a critical and contextual perspective is essential in preparing PSTs to teach diverse student populations (Pride, 2019). This becomes an especially relevant issue as classrooms become more racially, ethnically, culturally, and linguistically diverse, and almost concurrently, the proliferation of science and technology more intense. Little is known about how science might assist in addressing these concerns.

Despite an increasingly diverse social landscape of the United States (U.S. Census Bureau, 2015), and behind a veneer of tolerance and pluralism, many TEPs support a traditionalist model of schooling in which myths of meritocracy and colorblindness subsume the reality of limited educational opportunities for non-White students (Ash & Wiggan, 2018; Choi, 2008; Cochran-Smith, 1995; Croizet & Dutrévis, 2004; Darling-Hammond, 2005, 2010; Hickling-Hudson, 2004; Nembhard, 2005; Wiggan, 2007). In addition, TEPs may require only a single course on student diversity, which generally offers but a cursory examination of the various forces that have and continue to perpetuate racial inequality in education (Brown, 2007; Cochran-Smith, 2003; Hickling-Hudson, 2004). Given that the majority of the nation's teachers are White and middle-class and teach a growing cross section of the global population, coupled with uncritical assumptions of race, it is not unlikely that the resulting cultural mismatches between students and teachers elicit misconceptions of student ability (Emdin, 2016; Howard, 2006; Swartz, 2003).

Science Mis-Education

These cultural disconnections can be historically linked to social, political, and ideological myths and misconceptions used to stratify society based on race and ethnicity. Whereas Whites were traditionally regarded as the referential group, non-Whites were considered intellectually inferior and morally impoverished (Combe, 1830; Marks, 2008; Wiggan, 2007). The duplicitous use of science to justify racial discrimination along such lines that contributed to social stratification through programs like eugenics is still evident in education and society today.

For instance, the long-debunked notion of race as a "biological imperative" – whereby skin color is haphazardly linked to one's culture, or considered a determinant of intellectual capacity – remains embedded in schools and society, and receives the tacit endorsement of TEPs through the uncritical replication of dominant societal and cultural values (Bourdieu & Passeron, 1990; King, 1991; Mukhopadhyay & Henze, 2003; Smedley & Smedley, 2005). Such pejorative ways of thinking are most commonly promoted through the pervasive use of delimiting, implicitly racist achievement gap language uttered even among multicultural proponents (Hilliard, 2000; Ladson-Billings, 2006; Milner, 2012; Mukhopadhyay & Henze, 2003). Along with frequently negative media portrayal of non-Whites, White PSTs are encouraged to adopt an implicitly reified deficit model of their students based on racial and cultural stereotypes, in which success is "viewed as a deraced phenomenon achieved through meritocracy – if only individuals would try harder to do better" (Swartz, 2003, p. 256). The persistence of these cultural disconnections raises serious questions about the methods used in TEPs to prepare teachers for diverse settings.

Purpose Statement

Preparing students and teachers for the diverse tapestry of nations, ideas, and economies of a knowledge-driven and highly technical globalized world will invariably involve science (Kayumova, Avraamidou, & Adams, 2019). However, learning about and using science need not be confined to the science classroom, or as the means of production for vice and distraction from critical issues involving racial equality (Rodriguez & Morrison, 2019; Tolbert, Snook, Knox, & Udoinwang, 2016). Highly skilled but uncritical and historically myopic laborers will ultimately add little value to global society, and even less to themselves (Rahm & Brandt, 2016). The same being true for TEPs that do little to confront institutionalized forms of discrimination fixed by ideology, ignorance, or both (Cho & DeCastro-Ambrosetti, 2005; Howard, 2006; King, 1991; Swartz, 2003). Despite much new knowledge in the field of education about teaching diversity, as well as increased attention about socioscientific issues such as biodiversity conservation and climate change (Kayumova et al., 2019), little has been said about the utility of science, across disciplines, to establish the basis of cultural relevance through our common humanity (Amat, 2019; Banks, 2004; Bruna, 2016; Castro, 2010; Cochran-Smith & Zeichner, 2009; Gay, 2003; Kayumova et al., 2019; Milner, 2006; Sheets, 2005; van Dijk, 2013, 2014). This qualitative case study obtains data to help to address this research gap.

The purpose of this book is to investigate the role of science in teaching about diversity to undergraduate PSTs, as well as how and why science might be used to address issues of race, ethnicity, and culture in their TEPs. Overall, the goal is to determine the ability of science to deconstruct flawed narratives of race and human difference that have historically undermined racial parity in schools and society and legitimated pejorative assumptions of limited cognitive/academic ability of racial minority students. Three components are emphasized in this book. The primary objective is to explore the role of science in teaching about diversity in undergraduate preservice TEPs. Investigating this phenomenon involves analyzing multiple sources of evidence within a critical postmodern framework of science and diversity pedagogy. In addition, it aims to examine PST's perceptions of diversity before and after an instructional intervention to determine if, and/or to what extent, their views of diversity may have changed. Finally, using contemporary educational research on teacher preparation, it employs ongoing and concluding analyses of data to determine how PSTs are being prepared to teach in diverse settings, and specifically, the potential utility of science in this pursuit.

In this regard, this book advances that, when used as a catalyst for teaching about diversity, science can help unravel misconceptions about race by systematically engaging fundamental aspects of our shared humanity that evoke critical questions about racially discriminatory actions and discourse in schools and society. Embracing the notion of a common human origin of Africa would assist PSTs in creating more inclusive learning spaces, which has positive implications for student outcomes in the classroom and by extension, for humanity itself. Given the paucity of research on the use of science as a critical method for teaching diversity, this book is positioned to make an original contribution to a growing body of knowledge that examines the convergence of science, student diversity, and teacher education in the 21st century.

The Potential Role of Science

To inspire and to facilitate such critical analyses of diversity in teacher education, science must assume an alternate form: one that engages assumptions held by traditional science while forging a new path toward an antiracist teacher education. One fundamental assumption shared by scientists is that it can produce durable, though tentative scientific knowledge about the natural world. Not all questions can be answered through science, however, including concerns of personal and cultural beliefs or matters of good and evil. And yet, the ability to identify likely causal relationships using science helps provide context for such issues and in considering alternatives. In this sense, science can perform a vital function in education outside of the science classroom, as a critical response to the mis/undereducation of PSTs.

A significant aspect of scientific knowledge driving this investigation involves human genetic variation in DNA, or deoxyribonucleic acid. Namely, that the sum of genetic variation in DNA between humans is said to exist only within 1/100th of a percent. This means that out of the 25 thousand specific genes[7] and 3 billion base pairs of DNA that comprise the human genome, humans are still roughly 99.9% the same. The remaining 0.01% of human genetic variation is responsible for physical differences, such as blood type, skin color, or eye color, as well as gene mutations that denote susceptibility to certain diseases (National Human Genome Research Institute [NHGRI], 2006; Smedley & Smedley, 2005).

This suggests two important, interrelated points about science and human diversity that would be informative in PST education. The first and perhaps most obvious of these is that humans are far more similar than they are different; that we are divided only by various phenotypical differences and cultural nuance. Thus, variations in skin color and other superficial features reflect distinct genetic adaptations to particular geographical regions, not an inherent social hierarchy. This leads to the second point, which raises questions about social understandings of culture. PSTs would also do well to understand that culture is not something biologically intrinsic to humans, but rather, a series of customized rituals reproduced and transmitted generationally through language and particular forms of symbolic representation (Harris, 1999; Marks, 1995; Smedley & Smedley, 2005). Indeed, knowing that humans are 99.9% similar presents a somewhat glaring implication that many of the presumed differences between humans are less biological, and more so ideologies of race that support social and cultural dominance. Given that the reasons and methods for stratifying society along the lines of race or other such measures of "difference" amount to little more than social and cultural hubris, the development of greater critical awareness situated in our common humanity suggests that PSTs would be less susceptible to dysconscious racism in their future classrooms (King, 1991).

In this sense, the social construction of race, as a symbol of a cultural phenomenon of group dominance, is not an inherent feature of humanity itself (Donald, 1993; Winant, 2000). One implication being that social oppression is a complicit action, wherein the individual (inadvertently) participates in their subjugation by adopting a discourse designed to oppress (Apple, 2000; Foucault, 1977, 1980). And further, that learned behaviors of social and cultural dominance, whether professed as empirical truth, or otherwise operating as a tool for discrimination and racism, are alterable through better understandings of race and ethnicity using critical applications of science (Smedley & Smedley, 2005). Moreover, if social and cultural functions of a particular group are but external manifestations of that group, then progress is only as valuable as the level of critique within a society – the extent to which the beliefs and actions of the group are questioned from within. Without thorough and ongoing critical investigations of the past, the concept of race will remain a socially debilitating cultural phenomenon for what will no doubt be a globally interconnected posterity. As this book seeks to demonstrate, the depth and scope of diversity discourse in TEPs are radically limited by the absence of such discussions, and critical analyses of racial inequalities could, in fact, be assisted through the use of science.

In effect, science was positioned as a critical method for highlighting social, historical, and scientific inconsistencies and gaps of knowledge that willfully or otherwise perpetuates ideological myths and baseless legitimations for stratifying society. In addressing some of the inadequacies and related gaps in knowledge in teacher education, this book also seeks to contribute to the development of theories about critical science and diversity education by examining the transdisciplinary role and pedagogical impacts of science when used to teach about diversity in TEPs. Establishing a baseline for a critical dialog in education that begins with a common humanity makes possible more inclusive and affirming discussions of race and ethnicity. Using science to initiate such discussions therefore has the potential to strengthen TEPs and assist in preparing PSTs for effective performance in diverse classroom settings.

When used as an instrument for critical appraisals of social constructs such as race, science establishes a basis for antiracist teacher education and a means by which to instill critical awareness about such issues among future educators. A critical scientific approach that systematically examines and documents historical shifts in understandings of science and diversity can reveal how misguided ideological assumptions about human differences ultimately became a stratifying force in society. Further, using science to teach diversity in teacher education not only foregrounds a common humanity which supports racial equality but also addresses the growing need for a critically thinking and scientifically literate workforce. The dearth of research on this topic underscores the need for studies that examine the extent to which such an approach would lend practical assistance in TEPs as well as K-12 education. A natural progression toward this end could therefore include the CPSP here proposed.

Research Questions

An antiracist CPSP, which is discussed in further detail in the next chapter, seeks to position science as a critical method to deconstruct and then initiate a new discourse of diversity education that foregrounds a common humanity with only varied cultural lenses that form our respective worldviews such that new teachers might enter the classroom better prepared to reach students of all backgrounds. CPSP also supports several key features of the 21st-century classroom, such as higher-order critical thinking, scientific literacy, and collaboration. A specific focus of CPSP in this investigation is using science to initiate discussions on race and culture in ways that illuminate and more deeply probe issues of racial and ethnic inequality than multicultural or culturally responsive strategies alone, which often escape larger critical analyses of humanity/

human societies. An anticipated outcome is the development of a pedagogical method that supports effective teaching, and therefore, the students from diverse backgrounds that many PSTs will serve. Initial findings suggest that this method of teaching and inquiry into human differences has the potential of helping future teachers better understand and therefore reconcile misconceptions of race and ethnicity that tend to diminish the quality of instruction and suppress student achievement.

To further investigate this issue a single case study was conducted, in which three undergraduate PST education courses at an urban research university in the Southeastern United States serve as the unit of analysis. Using data collected from questionnaires, classroom observations, interviews, and responses during an instructional intervention, this book explores the utility of science as a critical method to teach about diversity. Along with ongoing data analyses, this book was guided by the following research questions:

1 *What is the role of science in teaching about diversity to undergraduate preservice teachers (at an urban research university in the Southeastern United States)?*
2 *How might this approach impact preservice teachers across disciplines in terms of raising greater critical awareness and influencing inclusive classroom practices?*

Context of Research: Case Study

The research questions were informed by the understanding that current demographic shifts among students do not culturally match the teacher workforce, and a theoretical framework that recognizes the need for a critical, interpretive, and incisive approach to discussing diversity issues such as race and ethnicity (Stake, 1995; Sheth, 2018). Along with the theoretical framework, the research questions helped shape the chosen unit of analysis through which the phenomenon of the utility of science as a critical method for teaching about diversity to PSTs was investigated. To thoroughly examine this potential role of science in what is here referred to as CPSP, a single case study was conducted with three undergraduate courses designed for PSTs that address issues of diversity in education.

The research setting for this study was an urban research university located in a diverse metropolitan area in the Southeastern United States. The unit analysis in this study consisted of ten (10) students and two (2) instructors from three diversity-related courses designed for undergraduate PSTs. Multiple forms of data were collected, including field notes derived from classroom observations and demographic questionnaires with open-ended items about science and diversity, as well as in-depth interviews with students and professors. Questionnaires were administered following a researcher-class presentation and discussion. Along with permitting the use of multiple data points to generate rich descriptions, qualitative case study research was chosen for its ability to generate in-depth knowledge that through detailed examinations of "how" and "why" science might be useful for teaching about diversity will provide insight on this under-researched topic (Lincoln & Guba, 1985; Yin, 2012). The findings report analyses of these 12 participants' experiences with concepts of human diversity within CPSP strategies.

Role of the Researchers

As with any qualitative study, this research has a distinct ability to bring researcher(s)' first-hand perspectives to enhance understandings of complex social phenomena (Glesne, 2011). As researchers, we assert that our subjective views and theoretical positioning are also inextricably embedded within qualitative investigations. As such, attempts should be made to bracket, or systematically account for, such views in the research process, particularly for studies that involve direct interaction with human subjects (Creswell, 2007; Kvale, 2007; Peshkin, 1988; Rubin & Rubin, 2005). Given this inextricable link between researchers and research, in outlining the methodological context of this study on science, diversity, and education, the researchers' experiences as often doubly marginalized (Black-White biracial man, Black-Caribbean man, and Black woman) are important to note. Although we cannot be fully abstracted from the human experience, acknowledging influences linked to the role of the research can help mitigate subjectivities brought to the process.

Prior to this study, we taught science, technology, and/or history courses in K-12 education for several years. All of our experiences were in the southeastern United States, which included elementary, middle, and/or secondary levels. As postsecondary level, the researchers collectively taught undergraduate, graduate, and doctoral courses, some of which were at the proposed research site. Further, as a result of many real and perceived instances of marginalization and a subsequent desire to affect positive change as a critical factor in education, the apparent prevalence of uncritical dispositions about and preparation for diversity is of particular interest in this study. Owing to personal experiences as both members and educators of and for diverse groups, the researchers maintain a critical perspective of how knowledge and power are mediated and views issues in teacher education as being linked to power differentials in society. Thus, because it is largely through social institutions such as education that the ability to generate self-directed meanings is governed, and whereby the legitimacy of knowledge is most noticeably brokered and deemed either valid or invalid relative to the dominant group (Altbach, 1971; Apple, 1978; Blake & Masschelein, 2003; Blumer, 1969; deMarrais & LeCompte, 1998; Gandin, 2011). Thus, as critical researchers and students/teachers of education, it is the researchers' position that there exist both hidden and explicit (social) oppressions functionally mediated through classroom instruction, and that the effects of which continually shape individuals' social and educational experiences. This perspective not only supported the epistemological and theoretical basis for the book, but also offered an analytical and interpretive lens capable of piercing the surface of diversity discourse in teacher education during the investigation. The overall aim was to uncover that which might otherwise reduce PST training to the reproduction dominant cultural values and social hierarchies (Aronowitz & Giroux, 1991; Bourdieu & Passeron, 1990; deMarrais & LeCompte, 1998).

Research Context

This context for this book was influenced by the researchers' role as a minority and students/teachers of education, who critically view society as being imbued with oppressive discourses and discriminatory treatment of traditionally marginalized racial and ethnic groups. In addition, prior examinations of literature on science, diversity, and education may have also affected how diversity issues of race and ethnicity in education were approached throughout the study. In this sense, the collection and ongoing analyses of questionnaire, observation, and interview data were likely influenced by the personal and professional experiences of the researchers. However, the continued development of personal and academic understandings of what it means to be relegated to the fringe of society and its effects in the classroom also provided an acute sensitivity for issues involving cultural mismatches between students and teachers, and importantly, the significance of being prepared to teach in diverse settings.

This book captures both structural and agentic qualities[8] regarding PSTs, TEPs, and the understudied phenomenon of CPSP. As such, descriptive, as well as explanatory elements of case study research are addressed. Whereas descriptive questions address the "what" of social phenomena, explanatory questions might examine the "how" or "why" of these phenomena (Boeije, 2010, p. 25). The case for this research was framed within the context of one TEP in the college of education at an urban research university in the Southeast. This context defined the boundaries of the phenomenon under investigation – the role of science as a critical method for teaching about diversity. The unit of analysis through which this phenomenon was explored consisted of three PST education courses that address social issues in education such as student diversity and issues of race. An additional point of inquiry was considering how the use of science to illuminate issues of race and ethnicity might assist in preparing PSTs to teach in diverse classroom settings.

In all, the book is interested in the efficacy of science in teaching about diversity in TEPs, and how doing so would inform and promote more positive and inclusive teaching practices.

To that end, earlier discussed was the notion that while it cannot directly answer questions of a personal or cultural nature, science can provide likely causal relationships from which to draw conclusions about ethical issues (AAAS, 2010). In this sense, both the descriptive and explanatory components of this exploratory case study assisted in linking the collected data with participants' perceptions about using science as a critical method to discuss diversity issues. The central research question is descriptive to the extent that it can be answered with rich interpretations about the role of science in teaching about diversity, and exploratory in that it also proposes further inquiry along these lines (Yin, 2003). The secondary question is explanatory to the extent that the study seeks to explain how and why using science to teach about diversity would be beneficial to PSTs in particular, as well as TEPs in general. The context and participants, therefore, served as a means to describe and explain the phenomenon of using science to discuss diversity issues in teacher education, and thus to contribute knowledge about moving 21st–century teacher education toward positive, transformative and inclusive teacher practices.

Significance of the Work

Among the concerns of the United States in a global society is how to effectively educate an increasingly diverse cohort of students to develop a highly skilled, scientifically literate workforce (Priest, 2014). Banks (2006) states, for example, that the "deepening ethnic texture, interracial tension, and conflict and the increasing percentage of students who speak a first language other than English make multicultural education imperative in the 21st century" (p. xi). Given the exceeding complexity of the current geopolitical landscape, it is reasonable to suggest that the nation's ability to make sound, ethical decisions about the socioscientific issues influencing the course of humanity will necessarily rely on a diverse citizenry proficient in STEM disciplines such as science (Gilbert, 2013; Mansour & Wegerif, 2013a, 2013b; van Dijk, 2013, 2014). Multicultural education plays a key role in disentangling these concepts (Castagno, 2009, 2013), as many scholars contend our understandings of race, human difference, and Whiteness involve scientific knowledge and processes.

Demographic differences between student and teacher populations and the effects of these cultural mismatches in the classroom that impact this intended trajectory are well-documented (Cochran-Smith & Zeichner, 2009; Delpit, 2006; Emdin, 2016; Howard, 2006; Ladson-Billings, 1995b). These understandings have driven a significant amount of research on preparing PSTs to teach diverse student populations using multicultural and culturally responsive approaches (Barnes, 2006; Bennett, 2013; Castro, 2010; Gay, 2002; Huerta, 2020; Ladson-Billings, 2000; Milner, 2006; Sleeter, 2001). The absence of critical self-reflection among PSTs, however, continues to dampen efforts to develop positive and inclusive practices aimed at reaching students from diverse racial and ethnic backgrounds (Benton-Borghi & Chang, 2012; Gay, 2003; King, 1991). This lack of self-examination generally stems from and reinforces an all but imperceptible, internalized sense of privilege espoused by a traditionally dominant White culture (King, 1991; McIntosh, 1989). As a result, PST's misconceptions about the ability of students from racial and ethnic backgrounds different than their own continue to go largely unexamined in TEPs (Cochran-Smith & Zeichner, 2009; Howard, 2006).

Furthermore, in its best applications, science can improve the quality of life of humankind (Tan et al., 2012). Thus, in response to the growing need for scientifically literate citizens to sustain a knowledge-based society, an increasing number of studies have examined teacher preparation for diversity, though specifically, in science education (Atwater, 2011; Atwater, Russell, & Butler, 2013; Elmesky & Tobin, 2005; Jegede & Aikenhead, 1999; Le & Matias, 2018; Lee, 2001, 2011; Lee & Fradd, 1998; Mansour & Wegerif, 2013b; Tan et al., 2012). This research suggests that science proficiency and critical applications of scientific knowledge are essential not only for addressing present and future concerns related to science but also to support meaningful collaboration among diverse populations and viable participation needed to sustain a prosperous global democratic society (Le & Matias, 2018; Mansour & Wegerif, 2013a). Each of these areas of research offers insights for addressing 21st-century issues of science, (student) diversity, and (teacher) education. Yet, little is known about the role science might play in education outside of the science classroom, particularly as a critical, transdisciplinary method of preparing teachers for diverse settings. This research investigates this phenomenon using a case study design, including participant observations, interviews, content analysis of questionnaires, and an instructional intervention that examines diversity issues of race and ethnicity from a scientific perspective. And because the utility of science as a critical method for teaching about diversity is an understudied area, this case study research is uniquely positioned to contribute to the literature and knowledge base regarding 21st-century teacher preparation.

In some ways, the ambiguity illustrated by the particular social and scientific views of "diversity" in this book offers an alternate path for understanding student diversity. To that end, Castagno (2009) suggests that several common multicultural definitions and approaches are obscure and counterintuitive, often due to their elusive definitions and loosely defined goals. Additionally, some researchers suggest that it is imperative to disentangle concepts of Whiteness within multicultural efforts (Le & Matias, 2018). One in which a common humanity of cultural differences represents possibilities, rather than pathological ideologies of stratification (Wilber, 2000; Wilson, 1997). In other words, while distinct commonalities in the origins and processes of human activity may exist, the concept of diversity also represents an ability to adapt, and therefore promote the possibility of a more evolved discourse about race and ethnicity for a democratic society (Mansour & Wegerif, 2013b). Within this changeability and somewhat "discursive nature of nature" is the implication that a critical science of diversity can provide a bridge to understanding social and cultural differences in education (Wilson, 1998).

This research explores the boundary between the social and natural sciences by examining how science and culture interact – a challenging, though important endeavor for education reform. To the degree that human behavior is a cultural transmission and our understandings of which influenced by if not contingent upon insights afforded by science, the cultural histories of humanity will, likewise, remain deeply entwined with this notion (Wilson, 1998). As Wilson observes, any misunderstandings therein come not from the lack of natural confluence between them; rather, such "misunderstandings arise from ignorance of the terrain, not from a fundamental difference in mentality" (Wilson, 1998, p. 138). One implication is that a critical lens that explores the scientific and sociocultural (i.e., genetic/biological, anthropological, geographical) histories of the human species may well support greater understanding and empathy in contemporary education – perhaps toward what could be thought of as an "ecology of human equality."

Summary

In this introductory chapter, interrelated issues of science and diversity were presented within the larger context of 21st–century schools and society, which provided the basis for this research. The problem this book sought to address was then outlined, which included research related to science and student diversity that demonstrated the importance of science in society as well as the need for effective teacher preparation to support the development of a diverse and highly skilled future workforce. Also provided were the purpose of the book, the research questions guiding the study, and the study's significance. Given the profound implications of increasingly diverse schools with a homogenous, if uncritical, teaching force and the growing need for a scientifically literate citizenry, U.S. students' underwhelming science achievement and the ineffectual preparation of teachers for diverse settings were emphasized as areas of concern. To this end, a qualitative case study that investigates the critical application of science to teach about diversity was proposed as a means to a more democratic and sustainable future in a knowledge-based global society.

The overall structure of this book takes the form of six chapters, including this introductory chapter. Chapter 2 provides extant research that supports the driving inquiry and purpose of this research. It begins by outlining the theoretical dimensions and conceptual lens of CPSP, through which relevant literature is analyzed, as with the remainder of the study. Chapter 3 highlights the need for a new discourse of diversity in teacher education through the following topics: (a) science, diversity, and the sociopolitical; (b) multicultural and culturally responsive approaches to teaching about diversity and difference; and (c) the implications of using science as a critical method for teaching about diversity. In addition to providing context, this chapter reveals science as but an extension of society rather than the immutable monolith it is so often presumed to be. Only by humanizing it as such can science be put to purpose toward more critical and egalitarian ends in contemporary society and education.

Chapter 4 provides more descriptive information regarding the research context for this book. It explores the potential role of science in teaching about diversity and details the rationale for conducting the study. The following chapter outlines the framework, context, and literature that support the relevance and significance of this book.

Notes

1. This process began at the end of World War II in 1945. In particular, this process was triggered by the strategic advantage it afforded in the War and in ending the Space Race and the U.S. moon landing.
2. Before dropping again after 2003 to 85,000, with an additional 20,000 available for those foreigners with advanced degrees.
3. For example, telephones, radio, television, Internet.
4. The exponential presence of science in society is, arguably, more so a function of the unfinished project for techno-geopolitical dominance following the Second World War in 1945 (e.g., International Monetary Fund, World Bank, World Trade Organization to name a few).
5. For example, federal legislation such as NDEA; high-stakes testing/ accountability; and increased focus on URM in STEM.
6. 34 of which are affiliated with the Organisation for Economic Co-operation and Development (OECD).
7. See also *Human Genographic Project*; Spencer Wells.
8. Structure and agency conflicts/alignments in TEPs, with regard to teaching about diversity, etc.

2 Critical Postmodern Science Pedagogy

The goal of this book is to explore and thus better understand the potential role of science in teaching about diversity to undergraduate preservice teachers (PSTs). As mentioned, this research examines social and natural sciences and how science and culture interact. In doing so, this book considers ways science and science education can mediate educational reform. As such, the context provided in this chapter serves as the theoretical underpinning for discussions on race, identity, and multicultural education in science, which is discussed in more detail in later chapters.

An ancillary aim of this book is to determine how and why science might be used to address issues of race, ethnicity, and culture in teacher education programs (TEPs). Using elements of critical theory and postmodernism as a theoretical lens, this chapter reviews existing literature that addresses three key aspects of the book: (a) science, diversity, and the sociopolitical; (b) multicultural and culturally responsive approaches to teaching about diversity and difference; and (c) the implications of using science as a critical method for teaching about diversity. Particular attention is given to the exclusionary effects of discrimination, namely, the duplicitous use of science by dominant (race and gender) groups to subjugate diverse groups and individuals. The misguided assumptions of human difference are well documented throughout history and help to shape understandings about science's response to human diversity. Historical evidence of scientific misuse and misinformation has irreversibly exacerbated stratification in schools and society. Approached from a critical postmodern science pedagogy (CPSP) perspective, the sociohistorical context of science provides insights for TEPs.

Schools and society are growing more diverse, with longitudinal population predictions estimating that by 2060, minority students will reach 64% in the United States (U.S. Census Bureau, 2015). As such, the responsibility for preparing a democratic and critically thinking workforce falls largely on the nation's future educators. Compounding the difficulty of this task is a public education system in which the sordid legacy of racial discrimination and social exclusion in this country persist.

Investigations of urban schools and communities depict troubling evidence of structural inequities. Within these glaring inequities, minority students remain overrepresented in the number of dropouts, special education programs, and low education tracks (Balfanz, Byrnes, & Fox, 2012; Hilberth & Slate, 2014; Kozol, 2005; Schott Foundation, 2012). Much as they have since the 1970s, opportunity gaps and restrictions to social mobility resulting from racially motivated social dislocation, lack of (educational) resources, as well as low-quality instruction continue to disproportionately affect students in the nation's most underserved communities (Denton & Massey, 1993; Eggers & Massey, 1992; Kincheloe, 2010; Lipman, 2004; Massey & Eggers, 1990; Rebell & Wolff, 2008; Wacquant & Wilson, 1989; Wilson, 1991, 2010).

Over the last three decades, debates about effective teaching, as well as teaching in more culturally responsive and equitable ways, spurred from cultural mismatch and incongruence in the majority of classrooms. Here, cultural mismatch describes the juxtaposition of students from a diverse array of intersecting cultural and linguistic backgrounds with a culturally homogenous teaching force largely unprepared to teach diverse students (Brown, 2007; Castro, 2010; Howard, 2006; Milner, 2006). In this research, we posit that teachers' pedagogical content knowledge (PCK) and cultural awareness, both of themselves and the students in their classrooms, are indispensable for effective teaching and high levels of student achievement (Emdin, 2016; Gay, 2002; King, 1991; Ladson-Billings, 1995a; Swartz, 2003).

Given the national and international focus on scientific innovation for its role in addressing present and future societal concerns and the underrepresentation of racial and ethnic minorities in science education and science-related professions, one implication is that more discussions about diversity are needed in 21st-century science classrooms (National Association for Multicultural Education, 2019; National Science Teachers Association, 2020; Parsons & Carlone, 2013). In light of this, many of these discussions include filling the nation's STEM-gap by increasing the number of members from "underrepresented groups," such as women and racial minorities (Griffith, 2010; Hurtado et al., 2008, 2010; Lee, 2011; NRC, 2007a, 2010, 2011; National Science Board [NSB], 2007, 2010; Riegle-Crumb & King, 2010; Riegle-Crumb et al., 2011). Less often, but importantly, these discussions offer social critiques suggesting that preparing a diverse and scientifically literate workforce is achieved through greater awareness of the influences, and at times, the dominance of culture in science learning and teaching (Aikenhead, 1996, 1997; Atwater, 1989, 2011; Atwater & Riley, 1993; Atwater et al., 2013; Barton & Yang, 2000; Gilbert, 2013; Jegede & Aikenhead, 1999; Lee, 2001; Noblit, 2013). Virtually absent among these debates, however, are implications for the role of science in initiating meaningful discussions of human diversity toward inclusive and transformative ends, whether inside or outside of the science classroom. Recent research has explored the educational experiences of critical science pedagogy (Rahm & Brandt, 2016; Rodriguez, 2015), yet there is still a need to expand these studies to include more meaningful intersections of race (Kirmaci, Buxton, & Allexsaht-Snider, 2019). To assist in filling this gap in the knowledge base about science and diversity in education, this book explores the utility of science as a transdisciplinary, critical method to teach about diversity issues such as race, ethnicity, and culture in TEPs.

To situate the ensuing discussion, this chapter begins by presenting the critical postmodern framework through which issues of science and diversity in schools and society are examined. Next, the chapter presents a sociopolitical discussion on how science is often misaligned as a metanarrative to serve the interests of the dominant class. Next, the chapter then presents literature related to science and student diversity, including contributions, connections, and critiques of science and critical pedagogy, culturally responsive pedagogy (CRP), and multicultural education (MEd). The chapter concludes with a new direction that employs science as a critical postmodern method to teach about diversity, along with several strategies that demonstrate what this approach might look like. The overall aim of which is to explore the possibility of science to deconstruct flawed narratives of race and human difference that have historically compromised racial parity in schools and society.

Theoretical Framework

To better position science and human diversity within the context of 21st education, this section outlines what is at present a heuristic framework of CPSP, which draws from critical and postmodern theories to link science and diversity to their social and historical contingencies. The purpose of this framework is to assist in the deconstruction and interrogation of scientific knowledge, and in the critique of socially and racially discriminatory practices such as phrenology and eugenics that science has helped imbue within the public discourse (Parsons & Carlone, 2013). With the goal of human liberation, critical theory is a useful tool for questioning assumptions, as well as critical investigations of seemingly benign social and cultural patterns, which are typically deemed so if only because of their prevalence or normative role in the functioning of society (Horkheimer, 1982). On the other hand, postmodernism denotes a skeptical atomization of categories that dissolves absolutes, focusing instead on the particulars of a social and historical moment. This, in turn, helps gratify any contextual and conceptual gaps remaining from the broad, macrosocial perspective of societies and cultures afforded by critical perspectives, such as world-systems theory (Hopkins & Wallerstein, 1982; Horkheimer, 1982, 1993; Lyotard,

1984; Wallerstein, 1974, 2004). Thus, as Steinberg and Kincheloe (2010) contend:

> The synergy of the conversation between the postmodern critique and critical theory involves the interplay of the informed moral practice of criticality and the lenses of complexity of the counter-Cartesian domain. As it invokes its emancipatory system of meaning, critical theory provides postmodern modes of analysis with a normative grounding. (p. 140)

To this end, Steinberg and Kincheloe (2010) and Kincheloe and Berry (2004) promote the theoretical and methodological perspective of bricolage in educational research, which seeks to connect and extend the emancipatory project of critical theory toward the "egalitarian impulses of modernism" (Steinberg & Kincheloe, 2010, p. 141). Given the hybridity of cultures and styles in contemporary society, such reconceptualizations necessarily suggest a practical and theoretical extension to the postmodern (Best & Kellner, 1991, 2003). In this research, these concepts are collectively applied in the examination of critical, culturally responsive, and multicultural forms of science pedagogy, and then deployed as a framework for exploring the role of science in teaching about diversity issues of race and ethnicity in PST education and its potential for raising awareness and promoting inclusive practices in TEPs.

Critical Theory

Critical theory is an invaluable theoretical perspective from which society can be examined for both hidden and overt oppressions enacted against any marginalized group, and through the social and cultural critiques it provides, supports the liberation of those enslaved by historical conditions arising out of capitalism (Horkheimer, 1982). Critical theory was popularized by a cadre of social and cultural theorists whose theoretical school of thought was commonly known as the Frankfurt School, founded in Germany in 1923 at Institute for Social Research (*Institut für Sozialforschung*), and included Herbert Marcuse, Theodor Adorno, Max Horkheimer, Friedrich Pollock, Leo Lowenthal, Erich Fromm, Walter Benjamin, and later, Jürgen Habermas (Agger, 1991, p. 107; Blake & Masschelein, 2003; McLaren, 2007, p. 185). The principal mission of these intellectuals was reconceptualizing Marxian class analyses to better align with the cultural changes in 20th-century society, ultimately beckoning a shift from the material contingencies of capitalism, toward social and cultural critiques of totalizing forms of alienation and "false consciousness" (Agger, 1991). Hence, a critical social theory is a theoretical framework that argues against all manifestations of false consciousness, which reifies the notion of an inevitable and rational social system at the expense of criticality in society (Agger, 1991; Alway, 1995; Blake & Masschelein, 2003; Horkheimer, 1982).

Horkheimer (1993) asserts that substantive critical social theories are those that emancipate the individual. This is achieved by explaining the shortcomings of the present and locating areas of social transformation and practical goals delineated by distinct analyses that investigate these particular foci. This suggests two requisite tasks of critical social analysis. First, the manifold nature of oppression suggests that, to some extent, relational and historical context is necessary for developing a critical theory framework. The purpose of overcoming this challenge is to avoid wholesale condemnation of a particular group or myopic calls for a panacea to heal society. Second, because oppression is not isolated to particular strata of society, critical social analyses must have the ability to operate at structural, cultural, and institutional levels, and with an interdisciplinary focus (Horkheimer, 1982, 1993). Moreover, critical social analyses merged with action define the basis from which critical theory seeks to liberate consciousness as well as the physical self from alienation and societal oppression. Frankfurt School theorists sought to achieve this through the mechanisms of capitalism, that is, to empower the human subject by way of interdisciplinary applications of information to create social consensus and return control over the democratic process in a capitalist society to the individual (Horkheimer, 1982, 1993). In this sense, analyses that employ critical theory are focused on raising critical awareness of how power differentials manifest in society as well as an individual's tacit or direct complicity in the oppression of others, such as the use of race as an instrument of stratification in society.

The main political thrust of the Frankfurt School was to undermine the positivist science and the ideologies of reason that stemmed from the Enlightenment through human emancipation, including deterministic Marxian views as well as those held by so-called traditional and social theorists such as Immanuel Kant, Auguste Comte, and Max Weber, by which the individual (proletariat laborer) was viewed as a passive rather than autonomous agent in society (Best & Kellner, 1991; Geuss, 1981; Horkheimer, 1982; Pleasants, 2002). This pursuit of critical theorists, argues Bohman (2005), marks the "horizon of critical social science" at which the "potentially self-defeating dialectic of freedom and power can be resolved ... in the continuing process of democratization, which in turn requires a fuller understanding of the requirements of freedom in institutions," though beyond its boundaries of hyper-rationalization of oppression (p. 354). As Blake and Masschelein (2003) note that in rejecting absolute empiricism and rationalism, critical theory begins to "question the transparency of society to the individual consciousness, and with it, the transparency of self to self" (p. 44). Alway (1995) adds that it is "a politics informed by a vision of 'distinctness without domination,' a politics of a plurality of agents, a multiplicity of actions, and a vastly expanded arena of political struggle" (p. 129).

In the context of this research, this suggests that to be an effective medium for change, a critical theory must therefore confront issues of race as they are observed during analyses of the social condition. And further that critical discussions about race can and should be had at the very sites where injustice occurs, such as TEPs and schools (King, 1991; McLaren & Farahmandpur, 2001; McLaren, Martin, Farahmandpur, & Jaramillo, 2004). A contemporary example of a critical theory driven by social justice and racial equality is the related tradition of critical race theory (CRT). CRT centralizes race in analyses of institutionalized discrimination and explores its effects on social relations (Delgado & Stefancic, 2001). In what began in the 1970s as a legal movement to expose juridical oppressions, often in the form of racially charged hominem arguments that disproportionately affected minorities, CRT would ultimately influence the field of education (Bell, 1995a; Delgado & Stefancic, 2001; Ladson-Billings, 1998; Tate, 1997; Wallace & Brand, 2012).

Not all Frankfurt School thinkers held the same view of the purpose and function of critical theory, however. Wiggershaus (1995) argues that there was never a unified critical theory originating from the Frankfurt School, but rather a series of related ideas centered around social and cultural critiques that borrowed from Marx's macrosocial economic analyses. He continues that, by the 1960s, the School had become an amalgam of the critical sociology of Adorno and Habermas (Adorno's research assistant), as well as earlier more radically oriented social and critical Marxist ideas as when the Institute was led by Max Horkheimer (Wiggershaus, 1995). Despite any internal transformations, the Frankfurt School largely remained true to its mission of extending the Marxian critique of industrial capitalism in applying it to the social and cultural nuance of contemporary society, yet keeping intact the notion of a pragmatic, analytical philosophy grounded in society by the lived experiences of the members within it (Agger, 1991; Wiggershaus, 1995). Among the more notable departures from the original conception of critical theory, however, was that of Habermas (1972, 1984, 1987), whose ideas mark the boundary between the first and second generations of the School (Geuss, 1981; McCarthy, 1985; Pleasants, 2002; Wiggershaus, 1995).

Critiques by Habermas (1984, 1987) about the direction of the Frankfurt School are significant to the extent that his theories on the role of language in critical thought served as the initial rupture from the critical theory put forth during the burgeoning years of the School (Agger, 1991; Pleasants, 2002). Believing that "Frankfurt School Critical Theory" had fallen victim to the metaphysical "paradigm of the philosophy of consciousness" and idealist conception of instrumental reason as had Marx's analysis of political economy, Habermas posited a paradigmatic shift toward a theory of communicative action in its place (Gunderson, 2014; Habermas, 1984, 1987; Pleasants, 2002). This marked shift included the somewhat structuralist-leaning notion that the rationality of communication was a more effective means for achieving liberation than was the instrumental rationality and conflicted relationship with the nature of first-generation critical theorists (Gunderson, 2014; Whitebook, 1979). And yet it was by this disengagement and foregrounding of language, as the means by which to disrupt manifestations of false consciousness, that further conceptualizing of critical theory was accomplished (Agger, 1991; Habermas, 1984, 1987; Held, 1980; Honneth, 1979; McCarthy, 1985).

In the years since this break with Frankfurt School, various applications of critical social theory have emerged, including world systems, postcolonial, and feminist permutations, many with pedagogical implications for multicultural education and science education (Capobianco, 2007; Carter, 2004, 2006; Feldman, 2001; Haraway, 1988, 1991, 2006; Hickling-Hudson, 2003; Johnson, Brown, Carlone, & Cuevas, 2011; Kellner, 1990; Sharpe, 2005; Wallerstein, 1974). As mentioned, this would increasingly include critical theories of social justice such as CRT, which sought to unveil racial inequality in the legal system through personal narratives that shifted the power back to the accused (Delgado, 1989). Though, as Kincheloe (2004) suggests, one of the earliest examples of critical theory scholarship aimed at "rending the veil" of race, as it were, was that of 19th-century urban sociologist W. E. B. Du Bois.

Evidenced by an intellectual legacy that includes such seminal works as *The Philadelphia Negro*, *The Souls of Black Folk*, and *The Conservation of Races*, Du Bois deftly confronted contrived, biological notions of race in 19th-century America from a social and historical perspective (Du Bois, 1897, 1899, 1903). In *The Conservation of Races*, for instance, Du Bois (1897) implicates the very foundation of science for endorsing "the grosser physical differences of color, hair and bone" that "go but a short way toward explaining the different roles which groups of men have played in Human Progress," rather than investigating the "differences – subtle, delicate and elusive, though they may be – which have silently but definitely separated men into groups" (p. 8). Then, further articulating this position, he states that:

> While these subtle forces have generally followed the natural cleavage of common blood, descent and physical peculiarities, they have at other times swept across and ignored these. At all times, however, they have divided human beings into races, which, while they perhaps transcend scientific definition, nevertheless, are clearly defined to the eye of the Historian and Sociologist. If this be true, then the history of the world is the history, not of individuals, but of groups, not of nations, but of races, and he who ignores or seeks to override the race idea in human history ignores and overrides the central

> thought of all history. What, then, is a race? It is a vast family of human beings, generally of common blood and language, always of common history, traditions and impulses, who are both voluntarily and involuntarily striving together for the accomplishment of certain more or less vividly conceived ideals of life. (Du Bois, 1897, p. 8)

Du Bois thus concludes that "[s]o far at least as intellectual and moral aptitudes are concerned we ought to speak of civilizations where we now speak of races" (Du Bois, 1911, p. 158). Above all, Du Bois makes clear the need for, and indeed, the potential of applying critical perspectives in society, to the extent that his indefatigable criticality has informed present research in pursuit of uncovering the reprehensible effects of racism and discrimination on urban schools and communities (Hayes, 2006; Kincheloe, 2004).

These theoretical positions respectively highlight implications of a society in which the Subject moves ever closer to the emancipatory goals of critical social theory – to identify and confront institutionalized pathologies of social hierarchy as exercised through patriarchal capitalism and imperialist structuralism and contesting such manifestations of oppression in local and global societies (Feldman, 2001). This reflects what Wallerstein (1974) observes in his seminal work on world-systems theory, that the "ability to participate intelligently in the evolution of his own system is dependent on his ability to perceive the whole" (p. 10). Which is to say that the particular (human, idea, etc.) is indeed connected, but only ephemerally bound, and should therefore be viewed as both whole and constitutive; as both macro and microcontextual; interrelated yet distinct (Feldman, 2001).

Moreover, owing to its conflict and critical antecedents such as Marx and the Frankfurt School, the similarities as well as divergences of these contemporary derivatives point to the inherent interdisciplinary appeal of critical theory in that it offers:

> a multidisciplinary approach for social theory which combines perspectives drawn from political economy, sociology, cultural theory, philosophy, anthropology, and history. It thus overcomes the fragmentation endemic to established academic disciplines in order to address issues of broader interest. (Kellner, 1990, p. 12)

With respect to educational research, the above suggests that it is a worthy endeavor to conduct analyses that seek to disrupt concentrations of power and privilege to create more critical, egalitarian human relations. However, as Giroux (2004) and others have observed, the convergence of social, ideological, and technological forces in 21st-century society has created a variety of more obfuscated threats of oppression and cultural hegemony that require new methods and a new language with which to engage the struggle of liberation (Giroux, 2004; Kanpol & McLaren, 1995a).

As the influence and implications of science and diversity in 21st-century society continue to expand further afield, it becomes clearer that today's youth occupy an altogether different world than what existed even a few decades ago. This is evidenced not only by demographic and sociopolitical changes but in the way they can engage this new world through various information and communication technologies (ICTs) afforded by scientific innovation (Best & Kellner, 2003; Law, 2008). Likewise, technical rationalities now increasingly operate in plain view, within an ever-present neoliberal social scene driven by deregulated industry and free-market fundamentalism (Giroux, 2011; Giroux & Giroux, 2006; Rahm & Brandt, 2016).

Consider electronic devices such as computers, DVD players, and cellular telephones. Many have become staples in homes, schools, and businesses; some with access to a global communication network comprised of equal parts information, entertainment, and risk (Kahn & Kellner, 2003). Such innovations were made possible through scientific advances in ICT that produced the microprocessor[1] in the 1970s, which revolutionized the way digital bits of data were stored and retrieved (Noyce & Hoff, 1981). Microprocessors are small silicon chips containing "millions of transistors and other components that process millions of instructions per second" (Raymond, Kuyoro, Adekunle, & Awodele, 2013, p. 488). Due to their small size and ability to perform complex computing functions at rapid speeds, microprocessors can be found in numerous household and personal electronic devices, as well as industrial applications such as transportation, high-tech medical devices, and computer servers (Raymond et al., 2013).[2] And yet, despite their intended benefits, the dominant class uses some of the same products from these advances as social, political, and economic tools of control (Hall, 1999; Kahn & Kellner, 2003; Kellner, 2003). Take for instance the recent coronavirus (COVID-19) outbreak and the reliability of information and communication technology amid a global pandemic. Yet, the same telecommunication tools (i.e., internet) that enabled widespread interconnectivity and access were also the same mediums for misinformation, degradation, and vitriol (Yam, 2020).

In addition to social and political conversations, the ability of STEM to generate population control even has educational implications. For example, in a searing critique of high-stakes standardized testing and standardization, Lipman (2004) and Gilbert (2013) argue that the negative effects of school inequalities and standardized assessment have become deeply entrenched in public policy. Researchers' analysis of high-stakes assessments implicates the strict accountability measures and disciplinary tactics of *No Child Left Behind* (NCLB) Act of 2001, the revised *Every Student Succeeds Act* (ESSA) of 2015, and the *Race to the Top* (RttT) initiative of 2012 as major factors in the deskilling and disenfranchisement of teachers and administrators in underserved urban schools. Gilbert (2013) later critiques the role of standardized measures in capturing courageous and critical science pedagogy, considering much of which is commoditized for economic gain. Throughout the 21st century, state education departments and school districts not only saw an increased number of standardized assessments, but were also charged with collecting, managing, and reporting massive amounts of data, in the way of test scores and student demographic information (Sunderman & Orfield, 2006).

Despite NCLB's emphasis on school accountability and reform through state oversight, Sunderman and Orfield (2006) observed an overall dearth in professional and organizational infrastructure among state education departments in Arizona, California, Georgia, Illinois, New York, and Virginia, suggesting that not all "state agencies have the resources, knowledge, and organizational capacity to intervene on the scale demanded by NCLB" (p. 532). School districts in states lacking the financial, human, and technological resources (i.e., computers) to develop and implement the student information systems needed to meet the high data-reporting demand, faced deep interventions, and in some cases school closure (Sunderman & Orfield, 2006, pp. 547–548). Those that remained open did so under the panoptic threat of unemployment or further disenfranchisement, forcibly narrowing the curriculum to the most basic components required to pass the tests and discouraging teachers and administrators from critically engaging issues of race and class inequality and marginalization (Lipman, 2004, p. 140). Moreover, with respect to students' differential learning experiences and levels, prior exposure to the content on which they are assessed, Hilliard (2000) concludes that the perpetual focus on what are essentially invalid assessments, not only disempowers but conceals "a history of abuse of minorities and the poor" and continues to ignore the savage inequalities of schooling (p. 296).

Given the above, a critical perspective on the high-stakes era of education raises questions about the pervasive politics of science and social control in the form of an accountability regime obscured by data metrics that ostensibly support school reform (Hilliard, 2000; Lipman, 2004). Hilliard (2012) was led to consider whether standardized testing was part of "a quality control movement, as it is advertised, or … a decoy for something else" (p. 16). As it concerns this book, a critical perspective also draws our attention to broader themes regarding the role of science in schools and society. For instance, could it be applied as a critical method to teach about diversity? That is, could science be more than a techno-rational instrument of capitalism that transforms nature (students/teachers) into objects (data points) and divests human subjects of an inborn capacity and ontological vocation to be free – as in freedom (Aronowitz, 1988; Freire, 1970; Horkheimer & Adorno, 1972)?

A notable weakness of critical theory in this regard is its almost totemic posture, having a similar macrospective focus on the totality of society as had Marx in his economic class analyses (Mannheim, 1954).

Critical theories are at times wont to wholly engage the self-legitimating edifice of capitalism on the premise that while the new division of labor is deeply codified as monetary success and social status, it is little more than an offshoot of the well-devised stratification of industrial society, and with a similar visage of natural fact – presented as though alienation should be desired and/or that it has always been the case – hence, the need for ongoing discussions of false consciousness (Aronowitz, 1988). However rightful, such analyses are also wont to overlook the accompanying notion that just as the bureaucratic force of capitalism has grown increasingly complex alongside the development of science and technology, so too has the language and instruments of oppression, by capital or any other such mechanism, become increasingly evasive and euphemistic in its attempts to conceal the message of domination (Aronowitz, 1988; Giroux, 2004). One implication is that, as a direct by-product of the political economy of industrial modernism, critical theory alone is a rather cumbersome approach for engaging the rapidly changing social, scientific, and cultural backdrops of the 21st century that nearly demand contextualization, if they are to be acknowledged at all. Moreover, the oppressive advances of unfettered capitalism are no less stultifying in a global society; if anything, it has become more entwined with the notion of commonsense due in part to market-based social policies over the past several decades (Apple, 2010; Harvey, 2005; McLaren & Farahmandpur, 2001; Stiglitz, 2003). Thus, a Frankfurt School perspective might argue against any possibility of a reconciliation with science, even to teach diversity.

The class- and market-based critiques offered by Marx and extended into the personal domain of the human subject by critical social theory are no less applicable in the general sense (Pleasants, 2002). Despite the powerful additives it offers for pursuing human agency, however, because of its broad focus on historical materialism (i.e., the economic structures and processes by which human activity interacts with and is transformed by the social world) and subsequent lack of precision in the context of the particulars such as race – though, eventually, it would be brought to the fore by critical race theorists in legal studies (e.g., Bell, 1995a, 1995b; Delgado, 1989; Delgado & Stefancic, 2001) as well as in science education (e.g., Wallace & Brand, 2012) – critical theory is also constrained by myopia in the form of perpetual re/negotiation of social contracts in contemporary society (Kellner, 1990; Marcuse, 2009; McLaren et al., 2004).

The Unintended, Critical Role of Science

Importantly, in 1961, Marcuse (2001) alluded to the possibility that social change could occur in a technological society precisely *because* the material line between the bourgeoisie and proletariat was blurred when scientific determinism and technology improved the living conditions of citizens around the globe, thus delivering them to the site of resistance itself, as McLaren has often argued regarding education (Moraes & McLaren, 2003). In arguing that the impending universalization of the rational dominance of science could somehow create spaces for social change, Marcuse makes a significant, yet paradoxical observation; that increased technical rationality of dominant society actually provides the very opportunity for social change. Moreover, he posits that the probability of the working class will critique material inequities decreases proportionately with increases in technological improvement (e.g., medical advances since industrialism engendered more contact with dominant institutions, better health, and less alienation). As such, the social critique turns inward, forcing society to reexamine itself, as a collective, using the same instrumental mechanisms (i.e., technologies) initially produced to increase capital (Marcuse, 2001). The implication is that these instruments of scientific domination could, in fact, contribute to the decline of the dominant structures that produced them – a resistance from within (Kellner, 2000).

For example, youth culture generally excoriated for their lack of participation and engagement with the larger world. It is presumed that they are disengaged because they are apathetic, oppositional, or otherwise noncompliant (Best & Kellner, 2003; Giroux, 2011; Kincheloe, 2011a). However, to Marcuse (2001), it may be more appropriate to say that recent influxes of technology, consumerism, and other forms of neoliberal profiteering (i.e., rational dominance) have only desensitized youth and deepened their false consciousness. Thus, while even traditionally marginalized (race, class, gender) groups in the United States are compelled to purchase gadgets, such as computers, cell phones, televisions, and car accessories, their increasing desensitization is met by a critical awareness *within* the very confines of this fetishism of commodities (Kahn & Kellner, 2003; Kellner, 2003; Marcuse, 2001; Marx, 2009a, pp. 60–62).

Despite being extrinsically urged to keep pace with the changing world, they may eventually encounter a point in their desensitization, in a collective false consciousness, where technology can become the very tool they use to disrupt the hegemonic forces that created them. In one example, many recent uprisings in Arab states were organized online using social media such as Facebook and Twitter, where nearly 90% of Egyptian and Tunisian youth surveyed in March 2011 said that they used Facebook either to organize protests or raise public awareness about them (Huang, 2011). However, while social media is often criticized for operating as a façade for authentic and concrete advocacy, there are some promising results in relation to today's activism and social change movements. A recent poll by the *Pew Research Center* (2018) demonstrated that 67% of people in the United States expressed the importance of social media in invoking social change and as a basis for widespread justice movements. This is most recently demonstrated in international social efforts such as #MeToo, #BlackLivesMatter, and #LoveWins. Recent research from the United States, Spain, Turkey, and Ukraine suggests that "big data" released from Twitter, Facebook, and other social media enterprises offers "unprecedented opportunities" associated with "large-scale collective action and social movements" (Jost et al., 2018, p. 85). Notwithstanding, while there is much debate over the veracity of digital platforms, activism has undoubtedly shaped the multimedia landscape. This suggests, in part, technology's role in mobilizing people and engaging change.

Therefore, while what we "know" is largely determined by external/material structures controlled by dominant forces, Marcuse (2001) and McLaren (2001) suggest that when fetishism collapses, history and social consciousness folds onto itself in an awakening that reveals oppression (Hall, 1999; Kincheloe, 2008). Such that when understood in universal and in particular terms of and by the oppressed, they unite with each other, but more importantly, with themselves (rehumanized); this resistance from *within* rationally dominant structures lends insight for using science to teach about diversity (Kahn & Kellner, 2003; Kellner, 2003; Marcuse, 2001; McLaren, 2001). Notwithstanding, this book suggests that when guided by *critical* intentionality, science can have a similar impact on discussions of race and ethnicity in education. For instance, the advances of scientific knowledge about DNA and genetics help to liberate closeted discourse about human diversity. However, without a strong pluralistic context of the social and historical, discussions potentially lack critical interpretation. Without proper social and historical contextualization, a critical science for teaching diversity would lack the sufficient interpretive fluidity with which to engage complex intersections of race and ethnicity in teacher education (Bauman, 2000; Collins, 2000).

Postmodernism

It is in such spaces where the overlapping boundaries of labor divisions and social stratifications and in the interplay of worldviews that the role of postmodern can more clearly be seen as having a role beyond the aesthetic (Agger, 1991; Kellner, 1990). Yet to categorically define postmodernism is a task of futility; though it may suffice to refer to it as a series of overlapping sociohistorical moments, a postindustrial condition following World War II, concerned with social and cultural departures from a so-called modern, industrial society (Derrida, 2009; Foucault, 2009; Hartsock, 1989; Lyotard, 1984). Jameson (1998) offers that, at the very least, postmodernism is often socially and temporally positioned around this time, which perhaps lends credence to the prevalent notion of a post/modern binary. The term postmodernism has been applied in various cultural and aesthetic milieus over the last several decades, with an initial aim of blending and re/contextualizing modern elements of society into a more heterogeneous view of the world. In a general sense, Lyotard (1984) describes the postmodern as an "incredulity toward metanarratives," or skepticism of absolutes, and then, more specifically, he argues that "postmodern knowledge is not simply a tool of the authorities; it refines our sensitivity to differences and reinforces our ability to tolerate the incommensurable," and is thus a reaction to modernity and its project of self-legitimation(pp. xxiv–xxv). Many critiques using critical theory and postmodernism are fundamentally rooted in a similar discontent with the presumed certainty of logical positivism during the Enlightenment.

The Enlightenment marked what is referred to as the *humanistic* period, which lauded autonomous reason as the premise for social homogeneity and thus reductive claims of universalism. Embraced as such by these newly enlightened or "emancipated" individuals, a schism with the theological tradition of absolute authority seemed inevitable.[3] Soon after would follow increased global travel, the slave trade, and other forms of trade and domination. Such that by the 1800s, the superfluity of the global slave trade had given way to industrialism and capitalism, and with it, a spirited attempt to organize society, and hence, modernity (Weber, 2009; Williams, 1944). This was due in part to an ethos of individualism that threatened solidarity, but also the need to control modes of production and direct the flow of capital (Durkheim, 2009a, 2009b; Hartsock, 1989; Lyotard, 1984; Marx, 2009b; Morrison, 1995; Mouffe & Holdengräber, 1989). For Weber (2001, 2009), the efficiency of which necessitated a merger of the calculable science of rationalization with religion, as axiological rejoinders to industrial capitalism.

As Aronowitz (1988) notes, the "rise of Protestantism in leading industrializing countries in the eighteenth and nineteenth centuries seemed to provide moral sanction for the preeminent position of modern science as knowledge," whereby religious institutions were regarded as the ethical riposte to the secular embellishments of money and power in industrial society, albeit conjoined with state authority (p. 9). And so, whereas artisanal and mechanical modes of production – by which raw materials were transformed into finished products – gradually succumbed to engineering research and practices involving electronics, physics, and chemistry, scientific knowledge itself was institutionalized by the state as well as large corporations, thereby legitimating its role as the new order of "truth" in society (Aronowitz, 1988). Hence, science was not only "a hegemonic ideology of the new social order of capitalism and its industrial stage;" it was now "integrated into the practices and discourses of production" that animated both the Protestant work ethic and state-supported capitalism at large (Aronowitz, 1988, p. 9). This was made possible, in part, through the political economy of the natural sciences. Subsequently, the impact of this scientific and technological transference toward hyper-rationalization and commodification is etched into contemporary society, where information has become commodity (Marcuse, 2001; Rahm & Brandt, 2016). In this sense, to control data/information is to control capital. This is a departure from the physical alienation of industrial society, but similarly reliant on false consciousness and the pretense of a monocultural society to subsume the laborer/consumer (Lyotard, 2009; Marcuse, 2001).

With this context, a postmodern position is concerned with the cultural and sociohistorical conditions of postmodernity, where social, political, and cultural "norms" were increasingly dissolved in favor of a more pluralistic worldview (Agger, 1991; Crotty, 1998; Jameson, 1998; Nicol, 2009). Though not entirely divorced from the so-called modern period, postmodernism nevertheless reflects a distinct philosophical and cultural shift from the totalizing discourses (e.g., culture, art, and literature) of modern industrialized culture of efficiency to one of multiplicity and fluidity (Bauman, 2000; Beck, 1993). Put succinctly, the postmodern is a "disconcerting mixture of present, past, and future – or of cultural and spatial elements from these different times" (Lemert, 2009, p. 454). Moreover, and importantly, postmodernism rejects the "violent hierarchies" and "conceptual oppositions" assumed by the binary conflicts inherent within structuralism (e.g., language/text in Saussure), Marxism (e.g., bourgeois/proletariat), and abovementioned dialectics of critical theory, concerned as it is with subject/object and culture/nature distinctions (Aronowitz, 1988; Collins, 2000, p. 70; Derrida, 1978, 1982, pp. 41–42; Newman, 2001, p. 2). Such dichotomies imply that "[o]ne of the two terms governs the other (axiologically, logically, etc.), or has the upper hand" (Derrida, 1982, p. 41). As Collins (2000) explains, if "binary thinking … categorizes people, things, and ideas in terms of their difference from one another," then "[o]bjectification is central to this process of oppositional difference," and thus, a condition of domination (p. 70). Thus, it would appear that underlying the logic of binary oppositions there is an essentializing force (or center); a force which legitimates, as it creates, the very power structures of authority and oppression in society that critical theories seek to disinter, evident in the objectification of subordinated groups, such as the delimiting, bifurcated language of a Black/White achievement gap in education (Collins, 2000; Derrida, 1978, 1982).

For Derrida (1978) and Collins (2000), postmodernism does not presuppose a fixed center of objectivity from which such oppositional terms could even begin to produce stable meanings. A postmodern perspective seeks, therefore, neither to engage nor to replace these binaries as such, but rather, to deconstruct them; to overturn their hierarchical foundation; to displace them with a new concept "that can no longer be, and never could be, included in the previous regime" (Derrida, 1982, p. 41). For, to do otherwise, argues Derrida (1982), to omit this "phase" of deconstruction, is to also prevent "any means of *intervening* in the field effectively" (p. 41, original emphasis). And so, with no presumed "grand narrative" guiding this heterogeneous style bent on decentering metaphysical universals, all discourses are provided conceptual space for their own truths/reality, and are thus imbued with a certain relativism devoid of clear distinctions or answers (Crotty, 1998; Derrida, 1978, 1982; Foucault, 1980, 1997).

In this way, postmodernism provides us with the mechanism for a conceptual shift in thinking that moves beyond the constrained dialectics of critical theory, toward envisioning science as a contested discourse within matrices of social and historical forces rather than a dichotomous field of oppositions (Aronowitz, 1988). Though, given its traditional role as/within an institutionalized structure/discourse, Derrida (1978, p. 353) suggests that science must also "be thought of as a series of substitutions of center for center, as a linked chain of determinations of the center" ever seeking to reestablish itself as such (Derrida, 1982). Yet, as Newman (2001) observes, absent an "essential place of outside the system" from which to interrogate the "reflexive self-identity" of philosophical or scientific claims, like Derrida, one is forced to work "*within* the discourse of Western philosophy [and science] ... looking for hidden antagonisms that jeopardize," and thus, deconstruct its presumed authority (p. 2, original emphasis).

While this reliance on prevailing metaphysical discourse – to launch critiques within the very structures it aims to disassemble – is often construed as the fundamental deficiency of postmodernism, it also suggests the ability to entertain multiple perspectives of engages the notion of truth at a more fundamental level (Agger, 1991; Bohman, 2004). Multiple realities allow "truth" to be situated within a particular social and historical moment, and thus examined more contextually for how and why certain "truths" about race, embedded, for instance, as uncontested beliefs about Black inferiority/White superiority, are allowed to remain intact in schools and society, despite being toxic functions of "power" (Foucault, 1977; Hutchison, 2006; King, 1991, p. 187; Massey, 2004). Relatedly, for Agger (1991), postmodern social theory is generally conceived as having either an apologetic focus on social and cultural affectations of style and language, or a more critical bent that is decidedly more political. Critical postmodern social and educational theorists and researchers such as hooks (1990, 1994, 1999) and Kincheloe (2001, 2011a, 2011b) examine the social world through multiple lenses of marginality, including race, class, and gender and the kaleidoscope of intersecting oppressions that follow (Aronowitz & Giroux, 1991; Best & Kellner, 2003; Collins, 2000; Denzin, 1986; Giroux, 1991, 2004; Hartsock, 1989; hooks, 1990; Kilgore, 2004).

As later discussed, the postmodern position that there exists no absolute truth also helps explain the notion that "truth is what power does." Which is to say, truth itself is embedded in context and comes to be, or is granted, through consensus; it can and does only carry a particular meaning relative to the "truth-teller" and, in turn, their position of power and applications of said truth (Foucault, 1980). For instance, if a teacher accepts the unmerited and generally uncontested public consensus that non-White students are incapable of learning at high levels as true, then this "racialized truth" – as the inculcated function of a common social discourse of White superiority that internalized over time – regardless of its legitimacy as democratic truth, can produce a harmful reality with tangible implications for the self-image and future performance of such students (deMarrais & LeCompte, 1998; King, 1991). Thus, by the same notion, the very fact that there exists alternate versions of "reality" for PSTs, in which racial discrimination or biological determinism regarding one's intellectual capacity is, if only tacitly, accepted as just means for understanding students from diverse cultural backgrounds, suggests that to interrupt such misguided thinking necessitates an alternate truth/discourse socially and historically grounded in democratic freedom (Giroux, 2004; Mannheim, 1954).

In the interest of breaking this cycle, in *Ideology & Curriculum*, Apple (2004) employs a critical analytical method in social studies to demonstrate how power has been entrenched in the curriculum, with regard to topics such as trade unions. His critiques highlight mainstream attempts to defocus the curriculum and therefore elide contentious political issues, diffuses important discussions about the role of government in assisting underserved populations (Apple, 2004). As an analog to this and other critical approaches to teaching about diversity (e.g., hooks, 1994; McFalls & Cobb-Roberts, 2001; Orlowski, 2011), this book proposes using generally accepted truths of science to develop a critical, rigorous method and discourse with which to better understand and confront issues of human diversity in teacher education (Aronowitz, 1988). Kincheloe and Berry (2004) suggest such an approach might involve the multilayered perspective notion of *bricolage*. This research perspective involves context-specific interpretive inquiry that is connected to its participants and offers a multiparadigmatic approach and "a high-level cognitive process involving construction and reconstruction, contextual diagnosis, negotiation, and readjustment" (pp. 2–3). According to Kincheloe and Berry (2004), conceptualizing in research various domains of knowledge (e.g., philosophical, interpretive, narrative, semiotics, etc.) offers unlimited forms of discursive social, historical, cultural analyses, and poststructural critiques that can problematize otherwise monological research in multicultural, science, and teacher education (Castagno, 2009). Thus, using various elements of critical theory and postmodern inquiry, this book aims to apply such multiperspectival analyses to explore the role of science in teaching about diversity.

Critical Postmodern Science (Pedagogy)

Despite notions of hybridity or plurality in research, a critical de/constructive framework of science and diversity nevertheless implies a certain level of logical critique and analysis (Kincheloe & Berry, 2004; St. Pierre, 2000). In the natural and social sciences, this often takes on the form of *positivist objectivity*. Here, when contrasted with a postmodern perspective, positivist objectivity presumes that the influence of the observer and the observed can somehow be abstracted from an investigation of an issue or inquiry (Kvale, 2007; Patton, 1990). As Best and Kellner (1991) explain, postmodernism criticizes such positivist modernism "for its search for a foundation of knowledge, universalizing and totalizing claims, for its hubris to supply apodictic truth, and for its allegedly fallacious rationalism" (p. 4). As such, it is important to delineate the usage of empiricism and objectivity as it was implemented in this study.

Whither Objectivism in Critical Postmodern Science?

The principles that underpin various scientific and philosophical traditions are products of their own social and historical contexts; each generation learns from the myopia of the last. Regardless of intent, any social endeavor that involves human interaction is susceptible to external influence by that very fact. Through a lens of critical postmodern science, instead of the purely intractable nature of objectivity conceived of during the Enlightenment, or the constitutive essence envisioned in the philosophical tradition of transcendental phenomenology, scientific "knowledge" and "truth" are (should be) highly contested areas to be deconstructed (Aronowitz, 1988; Foucault, 1980). With regard to this ontological instability of epistemology, that is, the methods by which the truth of a thing is determined, Agger (1991) observes that a deconstructive science:

> ...continually raises its assumptions to full view and thus invites readers to join or challenge them...science no more than fiction can attain absolute truth, no matter how reflexive it is about its own values, assumptions, and methodological choices: Every deconstruction can be deconstructed. (p. 115)

In this sense, science is not consigned to a linear plane. Instead, it can be applied globally via the condition that it remains self-interrogatory – of itself (Aronowitz, 1988). Here, the logic of science proposed by Popper (1966, 2002) may be informative. Popper (2002) offers the notion of "intersubjective testing and criticism" to decenter the idea of "conviction" as a viable mode for establishing stable truth claims; he states that:

> a subjective experience, or a feeling of conviction, can never justify a scientific statement, and that within science it can play no part except that of an object of an empirical (a psychological) inquiry. No matter how intense a feeling of conviction it may be, it can never justify a statement. (p. 24)

To be objective in this sense, a "scientific statement" is never absolute to the extent that it must be intersubjectively testable; that it can infinitely undergo further testing on the empirical basis of this intersubjective testability, from which other testable statements can be deduced. One statement tests the next and so on. In other words, scientific knowledge is only stable insofar as it has reached that transitory precipice of intersubjective testability, upon which it rests in wait for the next empirical line of questioning to gauge the a/symmetry of its truth claims. To Popper (2002), this method of deductive reasoning is also the best alternative to inductive reasoning, primarily because it holds no singular statement to be true in the universal sense in the first place. From this perspective, if the essence of a person or thing did exist, it would be more likely to emerge under such circumstances. Given this notion of the empirical objectivity (i.e., intersubjective testability) of truth used in the ensuing discussion, it follows that any universal narrative, whether of an ideo/socio/teleo/logical, or otherwise political or scientific determinism, is (or should be) subjected to its own criticism if it is to contribute to the progress and not regress of democracy. This suggests a vision of science as the systematic process of ongoing self-interrogation of scientific knowledge and truth claims.

Of Truth: Power, Knowledge, and the (Self)Interrogation of Science

Foucault's (1980, 2002) writings on knowledge, truth, and power provide some insight in this regard. Consider the following, for example: "'Truth' is to be understood as a system of ordered procedures for the production, regulation, distribution, circulation and operation of statements" (Foucault, 1980, p. 130). In his observations, Foucault (1980) similarly suggests intersubjective testability of discourse by acknowledging the fragility of objectivity in such statements, evident in his claim regarding truth and power; that is, "what I am saying here is above all to be taken as a hypothesis ... however, I would like to put forward a few 'propositions' – not firm assertions, but simply suggestions to be further tested and evaluated" (pp. 132–133). Yet, like Foucault's words, what is expressed is often branded to one's likeness – as ideology, as "Truth" per se. However, for Foucault (1980) and Aronowitz (1988), "Truth" is a situated scientific discourse conferred upon within the social, political, and economic institutions that produce and support it (e.g., education, military, media, and technology). Moreover, as a "political economy of truth," it operates circularly within these systems of power, the effects of which extend into what Foucault (1980) refers to as a "regime of truth," directing the individual and society through tacit politics of truth; or the:

> types of discourse which it accepts and makes function as true; the mechanisms and instances which enable one to distinguish true and false statements, the means by which each is sanctioned; the techniques and procedures accorded value in the acquisition of truth; the status of those who are charged with saying what counts as true. (p. 131)

Though, to Foucault (1980), political problems arising out of the strained condition of scientific discourse and ideology are perhaps better understood in terms of "truth" and "power." Beyond critiquing ideological impositions on science and society, or aligning it with another that is perhaps more palatable, as Foucault (1980) argues, the political issue is not merely "changing people's consciousnesses … but the political, economic, institutional regime of the production of truth" (p. 133). This does not involve whimsical attempts to emancipate truth from systems of power, but in the detachment of the power of truth from hegemonic apparatuses and redirecting it toward more transformative ends; to wit: "[t]he political question … is not error, illusion, alienated consciousness or ideology; it is truth itself" (Foucault, 1980, p. 133).

In this sense, the organic researcher and teacher intellectual, who seeks neither to hold nor to profess universal "Truths," is therefore positioned to confront and dismantle social apparatuses of systemic oppression and related truth claims to power, by virtue of one's relative position within the institution itself (Fischman & McLaren, 2005; Giroux, 1988; Gramsci, 1971).[4] Even within the context of the classroom or scholarly inquiry, the potential of the teacher/researcher to exercise localized forms of power on the particular is always present, be it positive or negative. Thus, as Foucault (1980) warns, it would be a "dangerous error to discount [an individual] politically in [their] specific relation to a local form of power," particularly "on the grounds this is a specialist matter which doesn't concern the masses" (p. 131). A strong point made here by Foucault (1980) is that truth does not exist outside of power, nor does it lack in truth, as such. It is the *effects of power* that truth claims produce that should be of concern. And so, with respect to scientific or diversity discourse: truth is what power does.

Race scholars insightfully assert that even if unintentionally, when PSTs are unaware of the misleading "truths" about race produced by and exercised through the power of internalized disregard, derision, or deficit-based thinking about students from diverse backgrounds, it can have significant implications for self-efficacy and academic performance (Gay, 2002; Hickling-Hudson, 2003, 2004; Hilliard, 2000; Hutchison, 2009; Le & Matias, 2018; Rahm & Brandt, 2016). Viewed within a critical postmodern framework of science and diversity, here proposed as CPSP, such truth claims can be tested or falsified directly, through critical awareness and by purposeful positioning of oneself at the institutional levels of (teacher) education. Moreover, the effects of power and in/direct sequestration of knowledge can be disrupted in the interest of critical social transformation (Foucault, 1980, 2002; Marcuse, 2009; McLaren, 1993; McLaren & Farahmandpur, 2001; Popper, 1966, 2002). In the absence of such an ability to critique knowledge claims from these localized positions, the vacuum it leaves behind will almost certainly be filled with hollow "convictions" in the form of silent statements that indiscriminately oppress in the interest, rather than the critique of social power, yet to the detriment of society. In the deafening silence of such convictions, Nascimento (2007, p. 34) reminds us of both the imperative and obstacles for emancipation:

> To offer subordinated identity equal footing to compete in this democratic game, it is necessary to break the [effects of the] hegemony of the dominant identity, Eurocentric Whiteness, constructed so solidly and so well reinforced that it reigns silent without being perceived. (p. 34)

In this book, which aims to explore the role of science in teaching about diversity, the use of scientific knowledge is neither regarded nor presented as some new inflection of an already essentializing force (Gilbert, 2013). As it is in the scientific community, this approach seeks no absolute truth, but rather, truth through critically informed consensus, which can be contested when it no longer responds to a given set of circumstances (AAAS, 2010; Aronowitz, 1988; Popper, 1996, 2002). In this sense, was used as a critical and re/constructive method for teacher education with which to critique not only sociopolitical aspects of race and pluralism, operating under what McLaren (1997) refers to as the "discursive mantle of diversity," but also scientific discourse, and through this de/construction and decentralization, the oppressive truth claims about race and ethnicity, therein (McLaren, as cited in Nascimento, 2007, p. 34).

As earlier suggested, many truth claims about the intelligence and cognitive ability of minority students stem from unexamined truth claims made under the aegis of scientific knowledge (Durodoye, 2003; Elias & Feagin, 2016; Marks, 1995, 2008; Wiggan, 2007). While not all subscribe to these "truths," the emanation of such ideologies from institutional seats of power has no less allowed them to manifest in society, perforce, as "Truth." Drawing from a critical postmodern perspective of science, this book aims to unveil and decenter oppressive discourses by de/constructing new ways of using science in schools and society with special considerations for race and identity. In positing a common humanity that arose out of the African continent, it explores the notion of a CPSP of human diversity as a reflexive and transformative activity in which specious claims of science and society about diversity are interrogated, as well as the domain of science itself.

As documented in this chapter, extant research supports the driving inquiry and purpose of this book. This chapter begins by providing the theoretical dimensions and conceptual lens of CPSP, through which relevant literature was analyzed, as with the remainder of the study. Chapter 3 highlights the need for a new discourse of diversity in teacher education through the following topics: (a) science, diversity, and the sociopolitical; (b) multicultural and culturally responsive approaches to teaching about diversity and difference; and (c) the implications of using science as a critical method for teaching about diversity. In addition to providing context, Chapter 3 reveals science as but an extension of society, rather than the immutable monolith it is so often presumed to be. Only by humanizing it as such can science be put to purpose toward more critical and egalitarian ends in contemporary society and education. As mentioned, this chapter explains the inseparable connections of race and identity within science education, which is largely missing in current discourse.

Notes

1. "A microprocessor is usually a silicon chip that contains millions of transistors and other components that process millions of instructions per second integrated with memory chips and other special purpose chips, and directed by software" (Raymond et al., 2013, p. 488).
2. "Several items such as DVD players, cellular telephones, household appliances, car equipment, toys, light switches and dimmers, electrical circuit breakers, smoke alarms, battery packs, car keys, power tool and test instruments use microprocessors ... industrial items with microprocessors include: cars, boats, planes, trucks, heavy machinery, gasoline pumps, credit-card processing units, traffic control devices, elevators, computer servers, most high tech medical devices, digital kiosks, security systems, surveillance systems and even some doors with automatic entry" (Raymond et al., 2013, p. 490).
3. This refers to the 16th-century dissolution of the Catholic church authority, propelled by Protestant reformers such as Martin Luther and John Calvin.
4. As this can apply whomever, one need only to envision and execute.

3 Science, Diversity, and the Sociopolitical

Introduction

The conceptual lens of critical postmodern science pedagogy (CPSP) discussed in the previous chapter was used to examine educational literature that supports the driving inquiry and purpose of this book, which is to explore the role of science in teaching about diversity to preservice teachers (PSTs), as well as the potential of this CPSP to influence positive and inclusive teaching practices. In what follows, a brief social and historical context of science and diversity is provided through a CPSP lens, which illustrates some of the traditional misuses of institutionalized truth claims framed through dominant cultural ideology as absolute. Particular attention is given in this discussion to the exclusionary effects of racial discrimination through the duplicitous use of science by traditionally dominant groups to subjugate individuals based on phenotypic variations such as skin color. By documenting historical shifts in understandings about science and human diversity, misguided assumptions of human difference emerge that help explain how racial discrimination developed into such a pervasive and stratifying force in schools and society. From a critical postmodern scientific perspective, historical critiques of repressive ideologies embedded in science and society provide insights for teaching about diversity in teacher education programs (TEPs) toward more inclusive, positive, and emancipatory ends.

Ideology

The concept of "ideology" is deeply interred within sociopolitical discussions. It is as politically charged as it is perplexing, in that it carries a variety of assumptions and etymological baggage depending on its usage (Eagleton, 1991). Eagleton (1991) describes the word ideology as "a text, woven of a whole tissues of different conceptual strands … traced through by divergent histories," and offers several definitions to capture some of its myriad meanings:

- the process of production of meanings, signs, and values in social life;
- a body of ideas characteristic of a particular social group or class;
- ideas which help to legitimate a dominant political power;
- false ideas which help to legitimate a dominant political power;
- systematically distorted communication;
- that which offers a position for a subject;
- forms of thought motivated by social interests;
- identity thinking;
- socially necessary illusions;
- the conjuncture of discourse and power;
- the medium in which conscious social actors make sense of their world;
- action-oriented sets of beliefs;
- the confusion of linguistic and phenomenal reality;
- semiotic closure;
- the indispensable medium in which individuals live out their relations to a social structure;
- the process whereby social life is converted to a natural reality. (pp. 1–2)

Given the apparently ubiquitous yet contrary "ideations of ideology," it is important to outline more precisely the denotative assumptions operating in the present discussion, as well as the study itself (Eagleton, 1991). In all, the concept of this book is concerned with the notion of ideology in terms of "true and false cognition, with ideology as illusion, distortion and mystification," as well as "the function of ideas within social life" (Eagleton, 1991, p. 3). Moreover, with respect to what Mannheim (1954) suggests in his treatise on the sociology of knowledge are its local and global (i.e., micro- and macrolevel) dimensions of ideology – relative to its locus within society as well as that of the social subject, ideology has both particular and total forms.

From a critical postmodern perspective, ideology in the global sense is envisioned as an oppositional force in society-at-large that functions primarily to allay attempts to deconstruct and subvert authoritative power or to deter disaffected thinking among its members. This is achieved by way of imposing dominant cultural values, the result of which is often a false consciousness that inhabits the mind of uncritical subjects only to obscure the distinction between structure and agency, though, presumably, for some greater good (Apple, 1978; Kumashiro, 2008; Mannheim, 1954). Local ideologies point to individualized perspectives of social actors derived from their lived experiences that are, at once, obscured by illusions of autonomy within their respective conceptual and cultural schemas, while also tethered to a sense of allegiance if not reliance upon totalizing forms of ideology (Mannheim, 1954). In this way, personal ideologies are often skewed, if not sublimated, by the larger ideological mechanisms perpetuated by hegemonic cultural values/views in society, such as race, class, age, sex, and gender discrimination. This suggests that macroscopic (global, critical theory, society-based, and educational) and microscopic (localized, postmodern, schools, and classroom) aspects of positionality and respective influences of ideology on them have significant implications for teacher education, to which a multiperspectival lens such as CPSP could be applied to grasp and confront such issues more directly.

In other words, the pervasive nature of ideology in education is such that both the flow and the direction of (scientific) knowledge is inarguably influenced by social institutions and discourse, as well as the historical context in which they are situated (Foucault, 2002, 2003). For instance, while supported by the traditionally dominant cultural ideologies of Eurocentrism, deficit-based perceptions inhibit critical learning opportunities for the PST, and ultimately, the personal growth and academic performance of non-White students in their future classrooms. From this perspective, how and what knowledge is transmitted and received during formal and informal learning processes necessarily affects the re/formation and re/production of cultural values and identity in a society (Bourdieu & Passeron, 1990; deMarrais & LeCompte, 1998; Foucault, 2002, 2003). Whether through public debate, political platform, or media, and in nearly self-regulated fashion, such discursive practices form the parameters and referential measures by which value is ascribed to various forms of knowledge (Foucault, 1977, 1980, 2002).

Given that education is one of the most effective means to transmit (or deny) knowledge, racism and discrimination remain both socially and culturally situated within the social context of education, as well as reproduced through the language and processes of schooling (Altbach, 1971; Apple, 1978, 1993; deMarrais & LeCompte, 1998; Foucault, 1980). However, as Mannheim (1954) argues, in that "[t]he ideas expressed by the subject are thus regarded as functions of his existence," the onus for reconciling ideological oppositions falls back on the social actor (p. 50). This suggests that, with or without their awareness, PSTs are ultimately accountable for the ideological implications of conscious or dysconscious racism in the classroom (King, 1991). Inasmuch as the depth and scope of diversity discourse in TEPs are radically limited by this apparent absence of critical analyses of race and racial inequalities, CPSP is presented as a viable move toward more antiracist forms of teacher education. As such, this explores CPSP as a means to empower PSTs as critical social actors who by confronting misguided historical assertions about race and centralizing the notion of a common African origin, resist the false consciousness and compliance inherent in the silent, yet pernicious, discourse of student diversity in TEPs.

Omissions and Mis/Appropriations of Science

History provides several examples of attempts to legitimate discriminatory practices against non-White ethnic groups by appropriating the narratives of science and culture (Foeman, 2009). Lacking the epistemological variability of cultural influences, science tends to draw exclusively from a Eurocentric worldview. This view forms the basis of Western Modern Science (WMS) and creates a Eurocentric metanarrative of humanity that bends decisively toward that particular cultural group (Corsiglia & Snively, 2001; Quigley, 2009; Shizha, 2010). In the shadow of WMS, often conspicuously omitted from the discourse and tomes of historical and scientific knowledge are the many contributions of non-White cultures, such as the Ethiopian descendants of Ancient Egypt, or Kemet,[1] to science and medicine, as well as mathematics, philosophy, religion, and various aesthetics (Asante, 1990, 2000; Ben-Jochannan, 1973, 1991; Diop, 1974, 1981). One such omission is also a misattribution. While modern medicine is generally traced back to the Grecian, Hippocrates (c. 460 B.C.E. – c. 370 B.C.E.), the so-called father of medicine, its origins extend much further into the past, to his African predecessor and first scientist named Imhotep. Imhotep was a chief official, vizier, physician, as well as chief architect of the Step Pyramid at Saqqara under King Netjerykhet (Djoser), ruler of Ancient Egypt/Kemet during the 3rd Dynasty (c. 2980 B.C.E. – c. 2650 B.C.E.) (Asante, 2000; Brandt-Rauf & Brandt-Rauf, 1987; Diop, 1981; Hurry, 1926; Osler, 1921; The British Medical Journal, 1927).

The *Centre for Biomedical Egyptology* suggests that Kemetian (Egyptian) doctors were practicing credible medicine over a millennia before Hippocrates was born (Pharmacy and Medicine in Ancient Egypt, n.d.; Serageldin, 2013). Most interestingly, Hippocrates mentions Imhotep at the beginning of his Hippocratic Oath in 400 B.C.E., which is still used by doctors today, and honors the god Asclepius (Imhotep) (National Library of Medicine, 2002):

> I swear by Apollo the physician, and Asclepius, and Hygieia and Panacea and all the gods and goddesses as my witnesses, that, according to my ability and judgment, I will keep this Oath and this contract … (National Library of Medicine, 2002, para. 3)

Meanwhile, while Western science cites Hippocrates as the father of modern medicine, historical and anthropological research, on the other hand, demonstrates that Egyptians were the original archetypes (Diop, 1974; Hilliard, Williams, & Damali, 1987; James, 2010; Pharmacy and Medicine in Ancient Egypt, n.d.; Williams, 1987). This is an important recasting of African intellect and the diverse contributions to science (Figure 3.1).

Figure 3.1 Imhotep: From the Imhotep Museum, Saqqara, Egypt

Source: Imhotep: From the Imhotep Museum, Saqqara, Egypt.

Critical interrogations of Eurocentric world history reveal, for instance, early Ancient Egyptians most closely resemble "Black" individuals from Sub-Saharan Africa (Asante, 1990, 2000; Ben-Jochannan, 1991; Bernal, 1987; Brandt-Rauf & Brandt-Rauf, 1987; Diop, 1954, 1974, 1981; Hilliard, 1986, 1992; Hilliard et al., 1987; United Nations Educational, Scientific and Cultural Organization (UNESCO), 1974, 1978). In fact, primary sources from the ancient Greeks confirm these findings, as mentioned in Aristotle's *Physiognomonica* (350 B.C.E./1966) and Herodotus' *Histories* (440 B.C.E./2014). Both of their accounts describe the Egyptians with Black skin and woolly hair. Herodotus explains:

> For the people of Colchis are evidently Egyptian, and this I perceived for myself before I heard it from others. So when I had come to consider the matter I asked them both; and the Colchians had remembrance of the Egyptians more than the Egyptians of the Colchians; but the Egyptians said they believed that the Colchians were a portion of the army of Sesostris. That this was so I conjectured myself not only because they are dark-skinned and have curly hair (this of itself amounts to nothing, for there are other races which are so), but also still more because the Colchians, Egyptians, and Ethiopians alone of all the races of men have [practiced] circumcision from the first. The Phenicians and the Syrians who dwell in Palestine confess themselves that they have learnt it from the Egyptians, and the Syrians about the river Thermodon and the river Parthenios, and

the Macronians, who are their [neighbors], say that they have learnt it lately from the Colchians. (Herodotus, 440 B.C.E./2014, p. 104)

The above primary source account documents the importance of Egypt in history, as well as the fact that these were Black people. However, these topics are still perceived as a danger to Eurocentric values or, at best, contentious misalignments of history.

Notwithstanding, in relation to Egypt's relevance to science discussions, additional topics have been overlooked or removed from mainstream curricula. For example, Adams (1987) notes that up to 400,000 years ago *homo erectus* (later *homo sapiens*) understood concepts of space, time, and seasons. Additionally, Africans in the Nile-Valley region harvested, built granaries, and domesticated animals up to 11,000 years ago. Since Africans already had an understanding of the lunar calendar and crop cultivation (Marshack, 1972), this marked the initial beginnings of agricultural science (Adams, 1987). In relation to geological science, researchers found evidence of mining over 33,000 years ago (Adams, 1987). Ben-Jochannan (1997) in his work *Africa: Mother of Civilization* (1997), describes the evidence of copper and iron extraction from the Zimbabwean valley over 5,000 years ago. In relation to medicine, Majno's (1975) *The Healing Hand: Man and Wound in the Ancient World,* suggests prehistoric medicine began upwards of 30,000–400,000 years ago (Figure 3.2).

Figure 3.2 Imhotep – Father of Medicine: Imhotep Museum in Saqqara, Egypt

Source: Imhotep – Father of Medicine: Imhotep Museum in Saqqara, Egypt.

Without question, the Nile Valley has perhaps offered the most impressive STEM achievements in modern history. In addition to the incomparable monuments and statues that have defied scientists for millennia (Diop, 1974; Lumpkin, 1984), the evidence of Egypt's preeminent understandings of STEM still serves as the prototype for modern thought (Clarke, 1979; Diop, 1974; Hilliard et al., 1987; James, 2010; Rodney, 2011; Van Sertima, 1984). In relation to agricultural science, for example, hydraulic engineers in the First Dynasty, measured the height of Nile floodwaters during annual inundation (Adams, 1987). Adams (1987) further notes:

> Under the reign of the first pharaoh, Menes, also known as Narmer, hydraulicprojects dramatically increased .... The scope of hydraulic projects took on greater proportions with the beginning of pyramid-building around the Third Dynasty of Pharaoh Zoser .... These hydraulic projects were used to regulate the flood waters for transportation of construction materials for the pyramids of Pharaoh Khufu, Khafre, and Menkaure (Fourth Dynasty). (p. S-16)

In 1340 B.C.E., an Egyptian scientist developed hydraulic-lift device, known as the "shaduf," which raised water to higher elevations in order to cultivate land. Later in the Ptolemaic period, Kemetians invented "saqiya," which was the cattle or oxen driven devices designed to carry water (ibid). It is important to note that Archimedes "took this device back to Europe and, subsequently received credit for its development though it was being used in Egypt prior to his arrival there" (Adams, 1987, p. S-18). In relation to time and cosmology, Egyptians are well-documented for their preeminent grasp on seasons, weather, and time. In fact, Herodotus (440 B.C.E./2014) notes in his primary account:

> But as to those matters which concern men, the priests agreed with one another in saying that the Egyptians were the first of all men on earth to find out the course of the year, having divided the seasons into twelve parts to make up the hole; and this they said they found out from the stars: and they reckon to this extent more wisely than the Hellenes ... whereas the Egyptians, reckoning the twelve months at thirty days each, bring in also every year five days beyond the number, and thus circle of their seasons is completed and comes round to the same point whence it was set out. (p. 4)

Herodotus (440 B.C.E./2014), among others, notes the discovery of astronomy and mathematics in the Egyptian region.

In relation to biology and genealogy, Egyptologists confirm that the Shabaka Stone (25th Dynasty, 800 B.C.E.) showed evidence of species and evolution "at least 2,000 years before Charles Darwin developed his theory of species evolution" (Adams, 1987, p. S-19). Relatedly, in relation to medicine, Van Sertima (1984) confirms the use of vaccinations, autopsies, surgery, and the medicinal use of plants hundreds of years before Europeans colonized Egypt. As aforementioned, the Greek god Hippocrates was imitative of his predecessor, Imhotep (Asclepius). Imhotep is the Egyptian god still referenced today in the Hippocratic oath (National Library of Medicine, 2002; Pharmacy and Medicine in Ancient Egypt, n.d.; Serageldin, 2013).

Additionally, Egyptians mastered cosmology and astronomy. As early as 4000 B.C.E., ancient Kemetians (Egyptians) invented the "merket," which was a measuring instrument that measured sun shadows and served as a developing compass. As such, Egyptians later discovered astronomy, as they noticed, "five bright stars changing their positions among the fixed ones" (Adams, 1987, p. S-28). In relation to mathematics, Stecchini (1961, 1971) confirms, "all contemporary systems of measure are based on the standards established by the Egyptians" (Adams, 1987, p. S-21).

Additionally, several primary sources confirm the influence of Egyptian (Kemetian) science and mathematics on the Greco-Roman world. Plutarch's *Parallel Lives* (75 C.E./2012), Aristotle's *Metaphysics* (350 B.C.E./1966), and Plato's Laws (348 B.C.E./1968) each give attribute ancient Egyptians for being the predecessors of Greek and Roman thought. Plutarch's *Parallel Lives* describes the "wisdom of the Egyptians" (75 C.E./2012, p. 17), while Aristotle (350 B.C.E./1966) outlines how science and "mathematical arts" were "founded in Egypt" (p. 3). However, Plato (348 B.C.E./1968) provides a relevant excerpt for this discussion. In an interaction between Athenian and Clinias, Plato notes the importance of arithmetic and measurement for precolonial Egyptians. In the words of Athenian, Plato (348 B.C.E./1968) notes:

> One ought to declare, then, that the freeborn children should learn as much of these subjects as the innumerable crowd of children in Egypt learn along with their letters. First, as regards counting, lessons have been invented for the merest infants to learn, by way of play and fun – modes of dividing up apples and chaplets, so that the same totals are adjusted to larger and smaller groups, and modes of sorting out boxers and wrestlers, in byes and pairs, taking them alternately or consecutively, in their natural order. Moreover, by way of play, the teachers mix together bowls made of gold, bronze, silver and the like, and others distribute them, as I said, by groups of a single kind, adapting the rules of elementary arithmetic to play; and thus they are of service to the pupils for their future tasks of drilling, leading and marching armies, or of household management, and they render them both more helpful in every way to themselves

> and more alert. The next step of the teachers is to clear away, by lessons in weights and measures, a certain kind of ignorance, both absurd and disgraceful, which is naturally inherent in all men touching lines, surfaces and solids. (p. 819)

Yet, with all of the compounding proof that corroborates these historically accurate narratives, science discourse today continues to support the overt omission of African contributions. The next section examines the issue of systemic erasure and misappropriated commendation.

Historical Erasure

After the conquest of Kemet (Egypt) by "Alexander the Great" in 332 B.C.E., the Greeks looted and burned the Royal Library at Alexandria, which housed the teachings of African professors and philosophers (Hilliard et al., 1987; James, 2010; Williams, 1987). This library later became home to many of the well-known scholars of Grecian history. Clarke (1978, 1979) notes that a similar instance of African erasure occurred at the University of Sankore several centuries later in the 1600s. At the University of Sankore, Ahmed Baba was a well-known exceptional teacher and scholar. When describing Ahmed Baba, Clarke (1979) notes, "after his extradition from Timbuctoo his collection of 1,600 books, one the richest libraries of his day, was lost" (p. 36). This library was another epicenter of African intellectual history that captured research on science, philosophy, and African contributions, among other things. This event, similar to the earlier incident in Alexandria, documented the symbolic, yet visceral, erasure of African achievements from world history. Notwithstanding, there have been a few brave scholars, such as Cheikh Anta Diop, Chancellor Williams, and Théophile Obenga who have provided decolonial science history. While the breadth of their research is groundbreaking, much of it is unknown to the general public.

Cheikh Anta Diop's (1954, 1974, 1981) pioneering, multidisciplinary work, linked Ancient Egypt to Sub-Saharan African culture.[2] Much to the consternation of classical scholars (e.g., Lefkowitz, 1997), who prefer to centralize Europeans as the progenitors of and most significant contributors to civilization, critical applications of scientific inquiry and research rightly challenge such notions (Bernal, 1996). To that end, the use of scientific inquiry to critique scientific inquiry is undeviatingly significant for revealing commonly overlooked cultural contributions and historical facts.[3]

For instance, in 1974, Diop and colleague, Théophile Obenga, participated in a symposium hosted by UNESCO in Cairo, called "The Peopling of Ancient Egypt and the Deciphering of the Meriotic Script," held partly at Diop's urging. Transcripts show how scholars critical of Diop's work tirelessly attempted to upend his thoughtful erudition. The result of the three-day scholarly exercise was said to be inconclusive; although, the transcript suggests otherwise. It was clear that Diop was well prepared to defend his thesis, as he repeatedly justified his position, noting among other evidence the distinct physiognomy, art, racial, and linguistic characteristics that necessarily qualify Egypt as an African civilization. Interestingly, the symposium report concludes with a rather shocking statement of what amounts to forfeiture on behalf of the scholars who had for three days opposed Diop's work:

> Although the preparatory working paper ... gave particulars of what was desired, not all participants had prepared communications comparable with the painstakingly researched contributions of Professors Cheikh Anta Diop and Obenga. There was consequently a real lack of balance in the discussions. (UNESCO, 1974, p. 102)

The results of this symposium demonstrate the dynamism of ideology and its sociohistorical influence on science. Even when presenting historical facts acquired using scientific methods, it was unable to dissuade members of the scientific community from clinging to ideological convictions of European preeminence in the precolonial world. Mis/appropriations of this sort raise questions about how ideology became so entangled with science and surfaces problematic influences of the academy on pseudoscience and scientific racism.

Pseudoscience and Scientific Racism

In today's information-based age, diverse forms of knowledge are more easily collected and transmitted around the globe. However, the rapid proliferation of ideas and information has not been without conflict and distortion. Presently, as well as historically, the vestiges of hegemonic ideologies about race and ethnicity seek to sustain a metanarrative in which the values and appearance of one cultural group is privileged above all others. Consequently, many non-White cultures agonize through realities that deny them human rights and equality. One method of such distorting information to favor hegemonic versions of truth is through pseudoscience. More generally, this is accomplished by privileging scientific knowledge acquired through WMS over other forms of knowing. Pseudoscience refers to methodologically flawed practices or otherwise ineffectual logic that presumes scientific authority. Whereas, scientific racism proposes that the human species is naturally divided, and exists as such, based on an inherent gradation of human value. Since it emerged around the late-17th century, this line of pseudoscientific thinking has been used to maintain racial fallacies, and therefore, to justify racial inequality through misguided ideologies presented as scientific knowledge (Foeman, 2009; Marks, 2008).

By the 19th century, being propelled by the burgeoning force of industry and capital, science had solidly become an instrument of authority used to legitimate social stratification based on race and ethnicity. Among the remaining political bastions of White-hetero-patriarchy is the long-debunked argument of race as a biological imperative, as well as the pejorative "conflation of homosexuality and pathology" (Marks, 1996; Omi & Winant, 1994; Terry, 1997, p. 272; Witzig, 1996). Due to misguided assertions of genetic or phenotypic variability embodied by pseudoscientific racist practices such as phrenology and eugenics, science has since been instrumental in reinforcing dominant group ideological values and under the auspices of sound empirical research. For instance, 19th-century craniologists used the racially discriminatory practice of phrenology to classify morality and intelligence based on the features of an individuals' skull.[4] In effect, however, science was contorted through the prevailing cultural ideology of White supremacy into a form of pseudoscience by which the intelligence and "force of mental character" of non-Whites was erroneously held to be deficient, and the individuals themselves deemed inferior to Whites (Combe, 1830; Marks, 2008; Wiggan, 2007). Eugenics[5] doctrines of the 19th and 20th centuries similarly attempted to show a causal link between race and intelligence to support such claims (Herrnstein & Murray, 1994; Hilliard, 2000; Marks, 2008; Wiggan, 2007).

Fallaciously conflated with the evolutionary theories of species diversity postulated by Charles Darwin, as well as Gregor Mendel's genetics studies, eugenics became an increasingly popular, "scientifically credible" basis for social and racial stratification (Bouchard, 1988; Marks, 1996; Ordover, 2003). Though aimed at "national hygiene," and ostensibly verified by scientific studies as well as Francis Galton's mathematical explanations of intelligence, phrenology and eugenics did far more to legitimate racial, class, gender, and sexuality bias and exclusion (Collins, 1998; Marks, 2008; McKnight, 1997; Ordover, 2003; Proctor, 1988). As a result, countless generations have endured subjugation, discrimination, and miseducation, if only to redirect public discourse away from the development of critical consciousness to challenge the hardline views of sociopolitical and ideological fundamentalists that have traditionally been at odds with science and society about issues such as race, class, gender, and sexuality (Collins, 1998; Noble, 1993; Omi & Winant, 1994; Ordover, 2003; Rosario, 1997).

Pseudoscience versus Authentic Science

Despite masquerading as genuine scientific practice, such mis/applications of science that serve the interest of a few to the detriment of many reveal the insincere foundation of pseudoscience. Contrastingly, authentic science consists of inquiry-based practices in which *evidence* and *logic* are used to examine natural phenomena and develop comprehensive elucidations regarding their existence. This generally involves positioning scientific knowledge as the basis of everyday life, as demonstrated through real-world applications of scientific inquiry and problem-solving methods (Crawford, 2014; NRC, 2012). Using a CPSP lens, this book extends this notion of *authentic* science to include a more intricate arrangement of social and political issues. As such, CPSP aims to address scientific problems in local and global societies (Buxton, 2010; Eilks, Nielsen, & Hofstein, 2014). An authentic implementation of CPSP suggests the need to apply scientific knowledge toward the advancement of a diverse, democratic society and committed efforts to provide adequate healthcare, nutrition, and sanitary living conditions for *all* global citizens, not serve the interests of the select few.[6]

As demonstrated in previous sections, the inauthenticity of pseudoscience is often buttressed by an uncritical reliance on a fundamentally racist ideology. Popper's (1963) eventually favored lens of critical rationalism dispossesses the idea that knowledge itself and postulates it could ever be fully apprehended. Thus, the principal criterion for determining the true authenticity of a scientific theory is "its falsifiability, or refutability, or testability" (p. 37). This delineation is conceptualized as the extent to which a scientific hypothesis or theory can be systematically disproven. In this regard, to genuinely test a scientific theory should involve the willful attempt to falsify or refute the stability of the claim itself. Thus, as opposed to pseudo-rationalism and other logically untenable forms of ir/rational faith in reason that lack critical examination and comprehension,[7] Popper (1963, 1966) asserts that only evidence resulting from self-critical tests of refutability can help produce truly authentic scientific knowledge.

Kuhn (1962), on the other hand, argues that falsifiability is an incorrect choice. Kuhn asserts that scientific knowledge is possible through social discovery and a scientific consensus when new scientific ideas emerge. This view suggests that science is not concerned with falsification, but rather following a paradigm shift, from which revolutionary ideas that disrupt normative paradigms of the scientific knowledge it precedes (Kuhn, 1962). By decentering this normativity, authentic science is, in the Kuhnian sense, postmodern, in that it rejects the coherence and telos of fixity of normal or traditional science "in favor of multiplicity, plurality, fragmentation, and indeterminacy" (Best & Kellner, 1991, p. 5). At the outset, it is tethered to an incommensurable narrative of "normal science" striving to validate its own claims, only to become beleaguered by the drift, crisis, and revolution of discovery followed by a groundswell of scientific practice that initiates a paradigm shift toward new understandings (Kuhn, 1962). From a CPSP perspective, Kuhn's (1962) observations are exigent as it shows the existence of authentic science beyond the borders of the social institution of science and scientific research. This is an argument less evident in Popper's (1963) work until much later.

Further, and importantly, a CPSP perspective suggests that, in spite of any conflicts arising between Popper's insistence on refutability (or criticism) and Kuhn's paradigmatic pendulum of discovery, pseudoscience is rendered ineffectual by either conception of authentic science practice. For instance, a significant point of departure of pseudoscience from authentic science is marked by its non sequiturs and base attempts to reify the primacy of Whites at the expense of racial minorities, purportedly for the "greater good" of society. Here, both Popper's (1963) criterion of demarcation and Kuhn's (1962) scientific revolution threaten the pseudoscience of phrenology and eugenics which, above all, seek to escape any such refutation or change as it would only expose the inherently hollow bias of racial superiority upon which their claims rest.

In part, the response to such critical views of science and society, the argument of White superiority has moved from blatantly racist explanations of intelligence offered by Combe (1830) and other 19th-century pseudoscientists to more contemporary and sophisticated theses as that presented by Herrnstein and Murray (1994), which more or less masks a similar assertion using statistics and psychology. However, these untenable discussions do not go unaddressed. For instance, amid public debates surrounding Herrnstein and Murray's (1994) notion of an incidentally White, "cognitive elite" in *The Bell Curve*, which in many ways served to subtly reify the archaic notions that intelligence is somehow biologically determined by one's racial or ethnic identity, or cycle of inheritable poverty, the *American Anthropological Association* (AAA) put forth the following statement:

> WHEREAS all human beings are members of one species, *Homo sapiens*, and
>
> WHEREAS, differentiating species into biologically defined "races" has proven meaningless and unscientific as a way of explaining variation (whether in intelligence or other traits). (AAA, 1994)

Based on this statement, the AAA urges all stakeholders and citizens to understand the socially constructed concept of race and its consequences and to assure equal opportunity, as well as respect for diversity and in securing a harmonious quality of life for all of humanity (AAA, 1994).

Despite the scientific community's opposition against these vestiges of (scientific) racism and a host of scholars in education, sociology, and psychology who point out methodological and analytical flaws in studies such as Herrnstein and Murray's (1994), which was found to measure the amount of academic instruction one has received rather than inherent ability, such baseless arguments still exist, albeit discussed in more subtle and evasive ways than in centuries past (Devlin, Fienberg, Resnick, & Roeder, 1997; Fischer, 1996; Gould, 1996; Hilliard, 2000).[8]

These examples illustrate how science (read: pseudoscience) is embraced by the dominant class when it serves to legitimate and solidify traditional Eurocentric power structures. The ignoble history of such attempts to legitimate the superiority of Whites on the basis of pseudoscience demonstrates the depths to which scientific racism was rooted in the academy. In this sense, scientific racism is the function of reified ideological power structures in society, and the deepening false consciousness. Interestingly, when science is used to challenge those same ideological structures the institution resists. However, as Atwater (2010) suggests, each instance of suppressed scientific discovery and critical discourse begs the simple yet complex question of "who benefits?" Who benefits when critical appraisals of social institutions are silenced by the clamor of ideology? Who benefits when advancements facilitated by science that could otherwise offer equity in education and quality of life are suppressed? These are questions that (science) teachers should be prepared to ask to uncover racial discrimination in curricula and instructional practice (Atwater, 2010).

Next Steps

Given the sociopolitical context highlighted above, it is clear that scientific knowledge is highly implicated in the struggle for racial equality in the United States. Indeed, the social and historical contingencies of science illustrate the lack of the presumed value-neutrality of science, instead emphasizing that it is one of many narratives with particular truths about the natural world. Further, the suppression and instrumental role of science in shaping social institutions to secure positions of power for a few also suggest that the language and the application of scientific discourse is a contested domain (Foucault, 1980; Hodson, 1993, 1999). While, from a critical postmodern perspective, the ongoing interrogation of scientific knowledge and discourse in society supports the ability of science to provide context to our shared human experience; thus, in order to transcend itself, it must critique itself.

What follows is a critical analysis and synthesis of literature associated with the questions driving this investigation: (1) *What is the role of science in teaching about diversity, particularly in TEPs?* and (2) *How might such an approach impact PSTs across disciplines, in terms of raising greater critical awareness and influencing inclusive classroom practices?* However, research on using science to teach about diversity in educational settings is limited. The selected literature, therefore, addresses educational issues that focus on student diversity and their integration with science. The literature is organized into three major themes followed by the positioning of critical postmodern science pedagogy in the body of knowledge on teaching about diversity issues in the 21st century. It is important to again note that the use of the term "diversity" is here concerned with race, ethnicity and culture. Related issues of discrimination in society such as class, gender, sexuality, and language are also important aspects of human diversity and should be acknowledged as such. However, the latter issues are beyond the scope of the present discussion, which is specifically concerned with the role of science in teaching about race, ethnicity, and culture. Notwithstanding, we welcome further discourses that examine the role of the CPSP framework and gender, ability, and sexuality discussions (Boda, 2019; Giordano, 2017). Moreover, this new direction seeks to build on current research in science and student diversity literature.

Science, (Student) Diversity, and (Teacher) Education

During the same historical period as the ascendant role of scientific and technological innovation and diversification of society, social movements and educational reform efforts aimed at mitigating the effects of racial discrimination in schools and society were also emerging. Consistent with the ubiquitous presence of science, this section discusses key areas of research involving science, diversity, and education that address rapid demographic shifts and issues of equity and social justice over the past several decades. These interconnected topics are commonly discussed in the context of multicultural, culturally responsive (relevant or sustaining), and critical theories of education. This research focuses on conceptualizing methods that increase students' and teachers' critical epistemological awareness of diversity issues, such as race, class, gender, language, ability, and cultural issues, both in the classroom as well as in local/global societies (Abrams, Taylor, & Guo, 2013; Apple, 2010, 2011a, 2011b; Castro, 2010; Freire, 2005; Giroux, 1988; Shizha, 2010; Swartz, 2003).

The framework for CPSP buttresses its theoretical underpinnings using three major themes in the literature, each with centered foci on science, diversity, and education. First, the theme of critical (science) pedagogy finds connections to culturally responsive pedagogy and multicultural education (MuEd) (Cochran-Smith & Zeichner, 2009; Lee & Buxton, 2008). Critical perspectives in teacher education are categorically more political and tend to centralize concepts such as praxis, consciousness, agency, social justice, and transformation/emancipation, place (Atwater, 1996a, 1996b; Atwater & Riley, 1993; Barton & Yang, 2000; Buxton, 2010; Cochran-Smith, 2010; Hickling-Hudson, 2004; Kanpol & McLaren, 1995b; Ladson-Billings, 1995a; May, 1999; Tsurusaki, Barton, Tan, Koch, & Contento, 2013). Additionally, this research positions itself within two additional themes that address culturally responsive (science) pedagogy and multicultural (science) education. As such, there are new conceptualizations of learning and teaching diversity through science, including feminist, postcolonial, and postmodern critiques (Capobianco, 2007; Carter, 2004, 2006; Haraway, 1988; Hickling-Hudson, 2003; Sharpe, 2005). While these approaches may differ in scope and application, there is a collective commitment to concepts of social justice and empowerment, especially among racially and ethnically marginalized groups. As noted in CPSP, this is permissible by fostering critical and engaging preservice teacher education and science education programs that offer insights to teaching diversity.

Throughout this discussion, the notion of science having agency and responding to critical and pluralistic forms of science education and teacher education has implications beyond the science classroom. In this case, teaching diversity using science involves developing critical educators who are attentive to both student and community contexts (i.e., race, ethnicity, culture, etc.) as well as wider political implications of "mainstream" science (i.e., norms of science), and thus, continually seek to enhance curriculum and instruction in ways that are more responsive and inclusive (Barton & McLaren, 2001; Basu & Barton, 2010; Hodson, 1999). There is a lack of research on science as a critical method for teaching diversity to PSTs. Thus, this literature provides a basis from which to explore and analyze the implications of CPSP and thus respond to the following research questions: (1) *What is the role of science in teaching about diversity to undergraduate preservice teachers (at an urban research university in the Southeastern United States)?* and (2) *How might this approach impact preservice teachers across disciplines in terms of raising greater critical awareness and influencing inclusive classroom practices?*

Critical Pedagogy and Science

Critical pedagogy is often associated with themes of resistance via multicultural, culturally relevant/responsive, and antiracist approaches to teaching and learning science (Atwater, 2010, p. 20; Gill & Levidow, 1987; Hodson & Dennick, 1994). Additionally, critical pedagogy also approximates space and identity, given its focus on marginalized and underserved populations. Similar to the 20th-century progressive conceptualizations of an authentic, democratic, and constructivist education, critical pedagogy in science education presents knowledge as student-centered and co-constructed (by student and teacher), and poses problems in "real-world" contexts relative to the various cultural epistemologies brought into the classroom (Dewey, 1916; Freire, 1970, 2005; Popkewitz, 1998; Ültanir, 2012). The ultimate goal is to cultivate transformative intellectuals who possess the capacity to resist social oppressions of all kinds and entrench in the historical struggle for human agency through education, radical democracy, and student-centered dialogue (Giroux, 1988, 2005; Kanpol, 1992; Kanpol & McLaren, 1995b; McLaren, 1999). This requires a critical and emancipatory consciousness by which students, teachers, and educational researchers locate inequalities in society, and through these informed critiques assist in empowering humankind as social agents (Bandura, 1989, 1999, 2001; Canella, 2007; Freire, 1970; Kanpol & McLaren, 1995b).

From this perspective, critical science pedagogy suggests profound implications for unearthing latent forms of repression by confronting dominant narratives of racial stratification through scientific inquiry and investigation (Atwater, 1996b; Barton & McLaren, 2001; Basu & Barton, 2010). Rather than "banking" scientific information into students simply to assuage the 21st century demands of high-stakes assessments (Freire, 1970), critical science pedagogy instead emphasizes an ongoing dialogue that challenges hegemony. In fact, Freire's entire premise of critical pedagogy necessitates the presence of dialectic engagement (Carroll & Minkler, 2000; Kirmaci, Buxton, & Allexsaht-Snider, 2019; Miller, Brown, & Hopson, 2011).

Dialogue, in this sense, extends beyond insubstantial communication and instead promotes equality in the educational process (Carroll & Minkler, 2000). Freire problematizes Western educational systems that position teachers, not students, as experts. Instead, Freire promotes the reciprocal bidirectional relationship of "teacher-learners" and students become "learner teachers" (Carroll & Minkler, 2000). In this case, deconstructing student and teacher relationships can be a way of engaging effective dialogue and mediating power relationships in the classroom (Freire, 1970/2000). In doing so, critical science pedagogy frames the subject content *within* relevant social, cultural, and political contexts (Freire, 1970; Kirmaci et al., 2019; Lipman, 2004). It is important to note that subsequent critical pedagogical frameworks that focus on race and ethnicity also assert the importance of human agency and re-centering, which places people as agents of change rather than subjects (Asante, 1991).

When centralizing and empowering the student/subject within relevant contextualization, science is revealed not as indisputable, but rather as a fluid and contestable domain. However, correcting or "rewriting" social distortions that create material inequities in society relies on critical literacy, and as some scholars posit – scientific literacy (Priest, 2014; Roth & Lee, 2004). Within science literacy, it is important to consider the phenomenon of social construction. Studying social construction involves, examining how researchers produce and report data, and in turn, how the public uses the information to depict a social problem or issue (Himmelstein, 2014). These critical antagonisms, by design, are evidence of critical pedagogy (Freire, 2005), in that it describes a creative and political act of knowledge, or "an effort to read the world" in order to emerge as a counterhegemonic force to level hierarchies (Freire & Macedo, 1987, p. 29).

In a study on scientific discourse, Brown (2006) examined the role of language and how culturally and linguistically diverse people could gain from clearer understandings of what is typically considered an *unfamiliar* enterprise. Brown (2006) contested that "the highly political nature of public discourse and its connection to student identity presented students with the challenge of balancing their identity with their use of science discourse" (p. 121). Emdin (2008a, 2016) provides an example of confronting such issues in urban science education from a critical and culturally responsive perspective. His approach consists of co-generative dialogue through popular culture, as well as student autobiographies of their learning in which students were able to "read and write the world" through their own voices, using hip-hop and rap music (Emdin, 2007, 2008a, 2008b; Freire & Macedo, 1987). Emdin's (2007, 2008a, 2016) research positions students as co-creators and co-investigators of scientific knowledge, in that they are learning, speaking, and re/writing about scientific information as it applies to their personal lives, communities, or cultural backgrounds.

Emdin (2016) demonstrates how centering urban minority students within their own science knowledge construction promotes high-achievement and agency (Freire, 1970; Lee, 2001). An important aspect of a critical approach to science highlighted by Emdin (2008a) reframes the exclusively Eurocentric domains of science via the political act of using hip-hop to "own" the language or discourse of scientific thought (Aronowitz, 1988). Moreover, Emdin responds to researchers who find scientific discourse to be "a problematic component of science learning" (Brown, 2006), especially for culturally and linguistically diverse students who balance their rich, discursive identities with science discourse (p. 121). The emancipatory potential of science promotes the importance of teaching diversity, while also problematizing local and global issues of race and ethnicity.

From this perspective, science must extend beyond the high-stakes regime of accountability in the classroom (Blank, 2013; Lipman, 2004; Ritt, 2016). Roth and Lee (2004) contend, "[t]he pursuit of scientific literacy promoted by recent national agendas does little to address the diverse audiences, many of whom have been squeezed out of science in traditional approaches" (p. 6). Unfortunately, the pressures of high-stakes assessments have created a longitudinal decrease in science instruction over the last two decades (Blank, 2013). Recent research from the *Center of Educational Policy* demonstrated that 44% of school districts saw reduced instructional time in science, social studies, and the arts by an average of 2 hours and 40 minutes per week (Ritt, 2016). Of the teachers interviewed in the study, 75% cited high-stakes tests as the reason (ibid). Instead of the myopic focus on standardized assessments, science should instead be used to help students develop critical skills related to scientific knowledge by critically and politically engaging in relevant issues. Buxton (2010) suggests this might include community-oriented or critical scientific investigations, in which diverse urban students examine the differential effects of the quality, distribution, and consumption of natural resources, such as freshwater in the poor and affluent neighborhoods in their respective cities.

Similarly, other scholars have taken the same action-research approach to science disciplines and connected their findings to contemporary social issues like food access, health disparities, and environmental justice. For example, Akom (2011) researched food and natural resource inequities in relation to school desegregation in low-income communities, later coining the term "eco-apartheid," to capture the relationships between social epidemiology, environmental inequality, and structural racism. Additional research on environmental and scientific racism, for example, corrals the active collaborative between empiricism and activism. Recent events such as the water crisis in Flint, Michigan (Raphelson, 2017), climate change controversies (Funk & Hefferon, 2019; Kamarck, 2019), and the coronavirus (COVID-19) outbreak of 2019 (Center for Disease Control and Prevention, 2020; Human Rights Watch, 2020), for example, each levied partnerships between scientific communities, political agencies, and the general public. Such investigations are facilitated by critical discussions on the instrumental use of science to support capitalistic rather than human-centered agendas (Barton & McLaren, 2001).

In this sense, science education is not merely a classroom subject, but also a political and contestable arena that allows students to challenge distorted power differences. Additionally, critical science education fosters social change by eliminating the barriers and gatekeepers that distance science from students in diverse classrooms (Brown, 2006). Furthermore, adding critical approaches to science ensures less vacuous forms of MuEd and minimalizes essentializing forms of antiracism in science education.[9] Notwithstanding, in the larger context of science disciplines, the consistent application of such macrosocial inquiries is often difficult to achieve (Atwater, 2011; May, 1999; McLaren & Torres, 1999).

In the ruins of historical power differentials in society aimed at reproducing the cultural values of the dominant group, these forms of critical (science) pedagogy are characterized as combative and divisive tools of the political Left, or diluted forms of protesting that do little to problematize critical issues facing public school systems (Bowers, 2005, 2008). For instance, due to a supposed tendency toward utopian and positivist-normalizing forms of education, critics such as Bowers (2005) liken the ambition of critical pedagogues seeking to transform society with the market-based infusions into social policy promoted by the neoliberal project currently dominating social and economic policies around the globe. Further, Gur-Ze'ev (1998) asserts that "all current versions of critical pedagogy function as part and parcel of normalizing education and its violence" and alternatively suggests that a non-utopian, anti-positivist counter-resistance in science education would not privilege any universal or particular view of society so long as it is non-repressive and democratic (p. 463). Tubbs (2005), on the other hand, offers a rather different critique of critical pedagogy, as well as its postmodern permutations. Tubbs (2005) argues that despite its historical ties to critical theory, critical pedagogy tends to ignore philosophical self-examination, and have therefore lost sight of "the triadic relation of understanding, change, and objectivity" (p. 228). Here, attempts to unveil oppression rejects, or perhaps ignores, the role of abstract consciousness in self-critique (Rømer, 2011; Tubbs, 2005). Together, these critiques suggest that applications of critical pedagogy in science may require a more *responsive* and *self-critical* approach to prevent immobilizing the inherent transformative potential for teaching about diversity.

Culturally Responsive Pedagogy and Science

Culture offers indispensable epistemological and ontological understandings of self and others. Culturally responsive (or, relevant) pedagogy (CRP) suggests that classroom instruction influences cognition, and should be especially considered when teaching students from diverse ethnoracial, cultural, and linguistic traditions. The concept of culturally responsive (relevant) pedagogy was popularized by Ladson-Billings (1995a, 1995b, 2009) based on her experiences of the exemplary practices of educators who by centering instruction on their cultural identities, were able to promote high levels of minority achievement. Culturally responsive teachers understand that classroom content and skills are more interesting and meaningful when pedagogy is situated within students' "cultural and experiential filters" or lived experiences. In other words, culturally-responsive teachers understand the "dynamic or synergistic relationship between home/community culture and school culture," (Ladson-Billings, 1995b, p. 467) and use "the cultural characteristics, experiences, and perspectives of ethnically diverse students as conduits for teaching them more effectively," (Gay, 2002, p. 106) and to empower them by facilitating a critical consciousness of human agency (Ladson-Billings, 1995a, 1998). Students who are taught in responsive ways, irrespective of cultural background or academic discipline, tend to reach high levels of academic achievement (Gay, 2002; Ladson-Billings, 2009).

This position asserts that providing equitable access to education in general requires teachers who are both aware and receptive to the various nuances of culture in diverse classrooms. And further, that to do so they must also understand the value of inclusion in the classroom, as well as the importance of culture in how information is transmitted and received, which suggests that teachers must first attend to the beliefs and mis/conceptions they may hold about different races, ethnicities, and cultures. And while integrating students' cultures in the classroom can be challenging, Lee (2001) and Lee and Fradd (1998) contend that teachers with firm scientific knowledge, with respect to their depth of understanding of general science content and specific subject matter, as well as an understanding of language and culture often promote student success in the science classroom as well. In this way, culturally relevant science pedagogy proposes to engage minority youth, as well as the culturally unaware educator, in ways that increase science achievement and critical awareness of social justice issues of race and ethnicity by drawing attention to inequities within the communities where they teach and learn (Atwater et al., 2013; Barton, 2003; Barton & Berchini, 2013). Lee (2001) argues that this perspective also begs questions of "what counts as science, what should be taught, how science is taught, and how student learning can be assessed in valid and fair ways"; and further, that "the quality of educational experience suffers if Western science is imposed on students who do not share its system of meanings, symbols, and practices" (p. 499). As alluded by Buxton (2010) and Brown (2006), this ultimately suggests that equity of voice throughout scientific discourse, curricula, and critiques of scientific knowledge is of particular concern (Aronowitz, 1988; Atwater, 2010).

Together with multicultural science education, which is later discussed in more detail, critical and culturally responsive science pedagogy points to the need to decolonize science curricula, as well as provide entry points for students of diverse cultural backgrounds to learn and perform science (Aikenhead, 1996, 1997; Atwater, 1996b, 2010; Gay, 2002; Hickling-Hudson, 2003; Jegede & Aikenhead, 1999; Nieto & Bode, 2012). This includes developing the critical awareness of how dominant cultural groups have traditionally been privileged in STEM-related curricula "through the selective inclusion and exclusion of material" in various textbooks (Asante, 1991; Kumashiro, 2001, p. 4). Kumashiro (2001, p. 4) argues, for example, that "mathematics has been a tool of colonialism and imperialism" to dominate others by using the ability to quantify nature, time, space, and thus society, through an assumed "logic of control" and reason.

As to the subsequent inability of some students to connect to a Europeanized curriculum, Hickling-Hudson (2003) contends that this discourse of colonialism currently embedded in education is "culturally problematic," and "does emotional and intellectual violence" to those who do not assimilate to the westernized hegemonic program (p. 3). Swartz (2005) adds that the:

> dismissal and disregard for decades of scholarly research nestles close to the core meaning of censorship: the restriction of knowledge

> through a process that discriminates what may even enter the books selected as repositories of 'standard' knowledge by school systems. (p. 32)

The implication is that while science, and more generally, STEM achievement, are indeed important for future sustainability in this country, so too are the ramifications of unexamined science practice on local and global societies. Here, the effects of privileging certain forms of knowledge and diminishing the value of other cultures obscure issues related to the inequitable distribution of resources and perspectives. Lipman's (2000, 2002, 2004) studies demonstrate how this is especially true in urban schools and communities where *stakes is high*, to use a phrase from hip-hop artists De La Soul (1996). With respect to oppressive assessments that claim to "count ability" while claiming accountability they nevertheless rest under the guise and in the interest of political economy – some bullshit, really (Barton & McLaren, 2001; McLaren & Farahmandpur, 2001).

McLaren and Farahmandpur (2001) explain the relationships between corporate models of education and global economic policies of "advanced capital societies" (pp. 139–140). Their research asserts that decreased government funding has fostered a perturbing relationship between public education and private corporations, who seek to capitalize on a $5.5 trillion industry (Kamentez, 2016). Despite the aim of producing highly skilled workers through quality STEM education, under the corporate model of education, information has become the new commodity and is least accessible to the poorest students, many of whom are concentrated urban cities (Barton and McLaren, 2001; McLaren and Farahmandpur, 2001; Mackie, 2020). Once again, this was most notably with student technology inequities during the COVID-19 outbreak, which most impacted poor communities (Mackie, 2020). Thus, since urban cities tend to be centralized in STEM discussions, science education should therefore include culturally responsive perspectives and disentangle the implications of access and its delimiting manifestations in society (Darling-Hammond, 2005; Lipman, 2004; Nembhard, 2005).

Culturally responsive science pedagogy offers powerful alternatives to mainstream science teaching and learning by valuing culture and holding high standards, while re/emphasizing the need for consistent critical inquiry to examine how science is applied (Irvine, 1992; Ladson-Billings, 1995b; Nola & Irzık, 2005). In particular, involves forms of inquiry that examine micro- and macrostructural obstacles to science teaching and learning in a multicultural world and global political economy, as well as the ongoing critical analysis of human diversity. Doing so would provide a basis from which students could not only acquire scientific skills but become critical researchers in society. Though, as Mensah (2011) notes, "teaching in culturally relevant and or multicultural ways requires a knowledge base about teaching for cultural diversity as well as understanding the sociopolitical context of multicultural education" (p. 297).

Multicultural Education and Science

Following the *Brown Decision* in 1954 and Civil Rights Movement of the 1960s, concepts of multiculturalism and cultural pluralism became more prevalent in public and educational discourse and provided a curricular basis for MuEd (Banks, 1993a, 1993b; Banks & Banks, 2010; May, 1999; Sharpe, 2005). And indeed, over the course of its development, MuEd came to include more critical and culturally relevant pedagogical approaches; however, an early vision of MuEd was to:

> reform the school and other educational institutions so that students from diverse racial, ethnic, and social-class groups ... experience educational equality ... [and] to give both male and female students an equal chance to experience educational success and mobility. (Banks, 1993a, pp. 3–4)

The underlying hope for this new field was that it would provide a viable means to facilitate increased cultural intersections in society, as well as address issues of achievement disparities among minority students (Atwater, 2011; May, 1999). However, critics of MuEd contended that structural effects of racism were understated, and the curriculum, overemphasized (May, 1999). Antiracist criticisms, for example, included the notion that multicultural curricula were naïvely viewed as a panacea for increasing academic achievement and social mobility among minority students (Gill & Levidow, 1987; May, 1999; Swartz, 2005). Among these debates was the view that MuEd was but a neatly framed utopian treatment that was privileged at the expense of deeper analyses of power differentials and race relations in society, and that it lacked sufficient analytical depth and sociopolitical scope to adequately confront issues of racial discrimination that pervaded schools and society (May, 1999). To that end, since the 1990s, researchers have coined the term "critical multiculturalism" to encompass decolonial and anti-racist perspectives (Awad, 2011; Nieto & Bode, 2012).

In addition, there was a rather healthy skepticism about whether the dominant class would even allow MuEd to develop as it was proposed. Afrocentric scholars such as Molefi Asante, who being well-familiar with the history of suppressed knowledge and cultures of non-dominant groups, were rightly concerned with a hidden agenda of White multicultural proponents (Asante, 1991; Asante & Ravitch, 1991; Bernal, 1987; Diop, 1974; Swartz, 2005). Driving this concern was that notion that the multicultural movement would be co-opted and fashioned into a new form of Eurocentric hegemony, as was evident in The Freedman's Bureau's role in the mis/education of Blacks in the South during the mid-19th century (Anderson, 1988; Asante & Ravitch, 1991; Woodson, 1933). From this perspective, MuEd would amount to little more than a near-passive addition to the process of schooling, whose covert purpose would be to reinforce Eurocentric values rather than to critically intervene in the political, racialized debates and educational discourse in society (May, 1999).

Asante (1991, 2010) argues, without decentering Europe as the pinnacle of global achievement, MuEd does little to combat historical inaccuracy and marginalization. As such, projects such as the *Portland Baseline Essays* of 1987 attempted to address these inaccuracies by providing decentered multicultural curricula across academic subjects. The cultures represented in these essays included African, Asian, Latino/a, and Native American, which served as the first geocultural project of its kind. Public school districts in Atlanta, Milwaukee, and New York implemented the *Portland Baseline Essays* in the late 1980s and early 1990s (Binder, 2000; Leake & Leake, 1992). Notwithstanding, it faced challenges of consistent application as assimilationist narratives usurped the educational climate of the 1990s. As such, more traditional forms of multiculturalism were adopted and standardized.

In the mid-to-late 1990s, MuEd developed and refined its mission and as the field gained traction, various academic disciplines began incorporating the tenets of MuEd into the curriculum. Social studies were among the first domain-specific area to adopt MuEd as a curricular disposition, and science, among the last (Atwater, 2011). This inclusion of MuEd in science would be significant for years to come and increasingly move into the sociopolitical and social justice arenas (Atwater et al., 2013; Hickling-Hudson, 2004).

Throughout the 1990s and the 2000s, multicultural (science) education was increasingly informed by critical and culturally relevant pedagogical approaches. One intention of this infusion was to support a science education for diversity, in which MuEd practitioners connected student culture to school culture and helped galvanize the transformative project of critical pedagogy (Atwater, 1989, 1996b; Atwater & Riley, 1993; Atwater et al., 2013; Hurtado et al., 2008, 2010; Lee, 2011).

Multicultural Science Education

The growing diversity in this country has all but ensured that MuEd will remain a pillar of democratic education (Ball, 2006). Recent national attention regarding underrepresented racial and gender groups in STEM fields, as well as the need for a scientifically literate or "globally competitive" workforce, contributes to a growing body of literature on science education and diversity (Hurtado et al., 2008, 2010; NRC, 2007a,

2007b, 2007c, 2010; Riegle-Crumb et al., 2011; Wang, 2013). Since the late 1980s, MuEd has been at the forefront of such efforts to make classroom subjects, including science, more accessible. Thus, it has been more thoroughly conceptualized as multicultural science education (MuSE) (Atwater, 1989, 1996b; Atwater & Riley, 1993; Hodson, 1993).

In accordance with the missions of the *American Association for the Advancement of Science* (AAAS, 1989) and the National Standards for Science Education, outlined by the *National Research Council* (1996), and more recently the *National Science Teachers Association* (NSTA) (2020), and *National Association for Research in Science Teaching* (NARST) (2018), MuSE applies MuEd as a critical postmodern approach to promoting equitable teaching and learning in science (Atwater, 1996b). Here, MuSE assumes a postmodern stance against metanarratives inasmuch as science is conceptualized, both socially and culturally, as a constructed "language game" of power. Today's *Next Generation Science Standards'* (2013) purportedly inclusive language still fails to reflect the nation's growing diversity (Bruna, 2016). Thus, as consistently shown, science is a contested domain in which dominance and universalist presumptions can be interrogated, and the inclusion of autochthonous epistemologies in scientific inquiry can be explored (Aikenhead, 1996; Aronowitz, 1988; Atwater, 1989; Atwater & Riley, 1993; Carter, Larke, Singleton-Taylor, & Santos, 2003; Hickling-Hudson, 2003; Jegede & Aikenhead, 1999; Lee & Buxton, 2008; Lyotard, 2009).

Atwater (1996b) describes MuSE as "a field of inquiry with constructs, methodologies and processes aimed at providing equitable opportunities for all students to learn quality science," and suggests that multicultural science educators believe that:

> (1) all students can learn science, (2) every student is worthwhile to have in the science classrooms, and (3) cultural diversity is appreciated in science classrooms because it enhances rather than detracts from the richness and effectiveness of science learning. (pp. 34–35)

In this regard, MuSE scholarship demonstrates salient features aimed at improving science involvement and achievement among minority students. Among them is the mission to decenter the persistent view of science as a monolithic specialty devoid of culture, ethnicity, and gender, and only accessible to certain groups or individuals. In response to this, the National Association for Multicultural Education (2019) challenged educators to expand curricula to include more diverse perspectives. Extensive research has been conducted to examine the implications of the fact that science is still primarily taught from a male, Eurocentric epistemological perspective (Aikenhead, 1996, 1997, 1997; Atwater et al., 2013; Carter et al., 2003; Jegede & Aikenhead, 1999; Suriel & Atwater, 2012). Yet, due to the presumed neutrality of science, such embedded forms of racism are allowed to persist (Gill & Levidow, 1987).

In contrast, MuSE takes a social constructivist stance that seeks to give ownership of science learning to the students, while also emphasizing the significance and impact culture has on the learning process (Atwater, 1996b; Au, 1998; Lee, 2008; Sheets, 2005; Ültanir, 2012). To ensure this connection, various aspects of MuEd are implemented in "science curriculum, science teacher education, student science learning and instruction, science assessment, and science evaluation" (Atwater, 2010, p. 103). However, while multicultural learning activities are engaging and support democratic views of schools society, their implementation in science curricula has indeed proved rather challenging (Atwater, 2010; Banks & Banks, 2010). In addition, Rodriguez (1998) and Rivera Maulucci (2013) note that some PSTs will either directly or indirectly resist applications of critical pedagogy, culturally responsive pedagogy, or MuEd. From a CPSP perspective, this lack of responsiveness stems from the disassociation of social and historical conditions that circumvent the need for critical discussions of human diversity in education. Without deliberately unpacking complexities of race and ethnicity, the value of these approaches fades behind the moniker of "feel-good instruction" (Ladson-Billings, 1995a). Thus, in the absence of context regarding how and why race remains a significant point of contention in contemporary society (i.e., that/why racism still exists), there is also a risk of fomenting feelings of resentment among PSTs toward debates about such issues, thereby impeding the development of a critical conscious about race relations.

May (1999) and Nola and Irzık (2005) have argued that this is, in part, a result of the strong focus on curriculum integration that multicultural (science) education seems to conveniently elide critical inquiry in general, and deeper treatments of race and class in particular at the expense of PSTs. Moreover, similar to Asante's (1991) suspicions, McLaren and others have often argued that multicultural applications were little more than a *repackaging* of neoliberal market ideologies served as inclusivity or pluralism (Kanpol & McLaren, 1995b; McLaren & Torres, 1999). Despite its apparently heavy curricular focus or supposed capitulation to market-based ideologies of late capitalism proposed by McLaren and others, along with critical and culturally responsive insights, multicultural (science) education ultimately seeks to ensure that students of different cultural backgrounds can connect to academic content (Atwater, 1996b; McLaren & Torres, 1999). Furthermore, as Swartz (2005) argues, multicultural forms of education do not seek to be essentialist, or merely a substituted form of centrism. Instead, its aims are only to put into *clearer* context the contributions of diverse perspectives that reside outside (and within) the borders of European history and ideology. Asante (1991 notes that to expand beyond the realm of European ideology, which has monopolized scientific narratives for centuries, it is important to first recast history with historical accuracy. In this regard, educators should ultimately examine the role of power and marginality. For example, in a survey of U.S. textbooks, the Portland Baseline Essays (1987) notes:

> A whole continent (Africa) is deleted from the history of physics. In all 17 books surveyed, not once is a scientific discovery or discoverer identified as being of African origin. No Black scientist is pictured in any of the books, nor is a single Black scientist credited with any contribution. (Portland Baseline Essays, 1987, p. S-2)

Studies on science representation in 21st-century textbooks note these same discrepancies (Bruna, 2016; Hickman & Porfilio, 2012). It is important to realign the contributions of African-descent scientists and intellectuals with greater historical accuracy (Murfin, 1992). This is not to portray an essentialist view; rather, this helps to codify the gravity of pre-colonial African erasure from mainstream discourse. This is the essence of several decolonial frameworks and epistemologies, such as aboriginal and indigenous pedagogy, Chicano/a studies, and Afrocentricity, as mentioned (Asante, 1991; Huerta, 2020; Semali & Kincheloe, 1999). Considering the erroneous and falsified histories propagated in mainstream science curricula, this has sparked larger conversations regarding the need for de-colonial frameworks in U.S. schools (Asante, 1991; Huerta, 2020; Semali & Kincheloe, 1999; Watson & Wiggan, 2016). In response, this book challenges the presumed sterility of science, which often thwarts the possibilities of decentering Eurocentric perspectives.

Building on the approaches highlighted above, this book uses the multiperspectival lens of CPSP to explore the potential role of science in teaching about diversity and its implications for future practice. The critical framework of CPSP inherently challenges assumptions about race and status quo by examining it as a socially constructed concept of subjugation within the broader contexts of human history. In doing so, CPSP serves as an interpretive framework that *displaces* reified notions of biological determinism, human cognition, and student ability. This, in turn, allows for the collective reconstruction of human diversity to instead promote cultural multiplicity, which gives voice to silenced and marginalized students in 21st classrooms.

Toward a Critical Postmodern Science Pedagogy of Human Diversity

Along with multicultural and culturally relevant approaches to science, CPSP's focus on political equity and social justice allows for more substantive applications of MuSE. This not only helps situate MuSE within the context of the political economy, but also assists in the development of a more transformative agenda for science (McLaren & Torres, 1999). Critical multicultural and culturally responsive forms of science offer the ability to politicize macroscopic issues of class, gender, and socioeconomics, and to challenge revisionist forms of history that either distort or omit the many contributions of non-White cultures (Swartz, 2005). Embodied within these approaches is a respect for the localized, cultural streams of knowledge students bring into the classroom, including preferred discourses of science and identity (Brown, 2006; Emdin, 2016; Gay, 2002; Ladson-Billings, 1995b).

By expanding the discourse of science into teacher education, CPSP has the potential to enhance TEPs in general, as those who educate PSTs are uniquely positioned to develop critical researchers and agents of posterity. In this sense, critical researcher refers to teachers as change agents, who seek to empower students as well as to interrupt and transform social and political realities through education (Creswell, 2005, 2007). Here, CPSP makes use of what students can know about the natural world, while disallowing the adaptation of metanarratives and pathological hierarchies, such as racism and c/overt discrimination, to more aptly pursue a science of self-interrogation (Derrida, 1978, 1982; Lyotard, 1984; Wilber, 2000). Moreover, as it attends to the notion that we indeed generate a form of power through the local structures we inhabit, CPSP may be useful in helping PSTs develop more thorough understandings of issues involving race, ethnicity, and culture (Foucault, 1977, 1980). This book suggests that one possible approach for achieving this goal is the application of science as a critical method for teaching about diversity, wherein scientific knowledge and discourse are used to illuminate the notion that we share a common humanity that began in Africa but now comprised of irreducible threads of individual existence (Derrida, 1978, 1982; Diop, 1974). By deliberately turning our attention to the premise that we have more in common as humans than we do based on any cultural and phenotypic differences, a different picture of diversity may begin to emerge. Given the paucity of research on this subject, the following section provides a few strategies to help illustrate what this might look like in a teacher education course about diversity; however, the applications of such an approach are not limited to this area.

CPSP Strategies for Teaching Diversity

Critical postmodern science pedagogy antagonizes, interrupts, and enduringly questions the role of teaching and diversity. CPSP is a critical method for deconstructing common discourse about diversity and instead reconstructs humanity as having a common narrative viewed through a multiplicity of lenses. However, by no means is this approach envisioned as a magical antidote for redressing socially and historically embedded forms of discrimination. Discrimination and scientific bigotry are deeply woven into the fabric of public education in this country, as manifested by: the near-exclusive privileging of Eurocentric curricula, 21st-century school segregation, criminal subjugation via inequitable school discipline, academic tracking, and the insidious racism propagated by Black/White achievement gap rhetoric (Asante, 1991; Green, 1999; Hilliard, 2000; King, 1991; Kumashiro, 2001; Oakes, 2005). Therefore, it is important to note that the goal of CPSP is not to form yet another essentialist metanarrative. Nor is it the goal of CPSP to ignore the complexity of oppression. Instead, CPSP helps problematize the longitudinal trajectory of humanity to promote the importance of using science as a political tool of antiracism that assists in the development of critical researchers and educators for the 21st century.

The latent disapprobation of non-White students, along with the subsequently stolid diversity discourse in TEPs, ineffectively prepares science educators to address the growing needs of today's students. In this way, CPSP signifies a shift from redacted discussions on racial discrimination, toward more contextualized and critical debates about humanity that, as Geuss (1981) suggests, "makes the subjects in the society aware of their *own* origin" (p. 70, emphasis original). In contrast to the way that multicultural and culturally responsive approaches are generally taught in TEPs, which simply "enlighten the subjects about their own genesis or origin," or "explain to them how they became the subjects they are with the beliefs, attitudes, norms, etc. they have," CPSP seeks to unravel the web of discursive practices of racism in education in order to:

> [show] them under what conditions, in what 'context,' they acquired these beliefs, attitudes, and norms, and how they came to hold their basic world-picture, that is, how they came into being as social subjects. (Geuss, 1981, p. 70)

Yet, CPSP neither implies nor aims to provide an unequivocal rejoinder to endeavors of personal emancipation, as it is more the pursuit of critical reflexivity itself, and accordingly, the questions one is entreated to explore, from which salient insights about race and ethnicity might be acquired. The true potential of CPSP lies in its ability to support the development of critical consciousness about such issues with which PSTs might enter diverse classrooms better prepared to understand, reach, and thus effectively teach, each of their students. Moreover, an added benefit of this approach is that, regardless of discipline, it bespeaks the development of a critically thinking citizenry, thereby enhancing the potential of a democratic society (Mansour & Wegerif, 2013b).

Three strategies are here presented that, if used as supplements for diversity discussions with PSTs, could problematize the conflation of skin color with achievement and biologically deterministic misunderstandings of race, ethnicity, and culture (Gravlee, 2009; Smedley & Smedley, 2005). These strategies incorporate the scientific disciplines of biology, anthropology, genetics, and geography to highlight the notion of a common human origin in Africa, though not to replace one centrism with yet another. Rather, by reorienting humanity in this way through the use of recent scientific knowledge, these strategies become critical mechanisms through which PSTs might better understand their own identities, as an extension of the African Diaspora. A potential benefit of doing so might include the acknowledgment of a shared human connection between teachers and students. Additionally, it might help foster a greater propensity for PSTs to connect to students from backgrounds different than their own. Another potential benefit is that students and teachers with intercultural, indigenous, and hybrid dispositions may well find it easier to situate their identities and epistemologies in contemporary (science) classrooms after such hierarchical-leveling (Carter, 2004; Moosa-Mitha, 2005). Sadly, although these strategies are based on concepts readily found in K-16 science curricula, much of this information is either glossed over or simply forgotten. In any case, while these concepts do arise in some culturally relevant and MuSE literature (e.g., Atwater, 2010; Laughter & Adams, 2012), there is little evidence to suggest that such an approach has been used for teaching diversity to PSTs.

DNA ≠ Do Not Ask

One way CPSP can help initiate discussions about diversity is through a significant element of current scientific knowledge that involves the variation of human DNA. As noted in Chapter 1, the sum of genetic variation between humans exists only within 1/100th of a percent (0.01%), which means that we share 99.9% of our DNA sequence. This is especially valuable when considering the preponderance of information that amplifies and reinforces synthetic relationships between (and within) phenotypic differences, such as skin color. Likewise, this information illuminates two important points about science and human diversity. The first, and perhaps, most obvious of these points is that humans have far more similarities than differences. So much of our social life is spent incessantly comparing ourselves one with another, though with the rather simple goal of validation within our human interactions (Blumer, 1969). Thus, to one degree or another, we tend to seek this validation extrinsically. By understanding that the genetic variability between individuals is so slim – something that is often taught in high school and college biology courses – in the context of education, we are encouraged to see that differences such as race are based on social constructions and are therefore malleable rather than immutable (Foeman, 2009; Omi & Winant, 1994; Pollock, 2006; Smedley & Smedley, 2005; Winant, 2000). This presents opportunities for changing the way we view ourselves in relation to others by understanding that no one "race" is better or worse than the other, only different – and not by much at that.

The second, interrelated point is perhaps more of an indictment. Given that so much of the global economic sustainability of posterity rests on knowledge, application and innovation of science, it would seem prudent to acknowledge the significance of our similarities. The fact that humans are 99.9% similar presents glaring implications that many of the presumed differences between humans are not at all biological, but rather ideologically based on race and hegemony for the purpose of social and cultural dominance. Presenting this scientific discovery in an education class has the potential of developing students' critical consciousness about racial formation, and subsequent stratification, in the United States (Foeman, 2009; Mukhopadhyay & Henze, 2003; Smedley & Smedley, 2005). Thus, if racial stratification is but a function of social and cultural hubris, then appealing to PST's critical consciousness might reveal a common humanity within each person and student. Here, human history and connection should be valued firstly on the basis of a common African ancestry.

Pangaea

Another way to stimulate discussions about diversity using CPSP is through a brief re/teaching about the once supercontinent of Pangaea, an interconnected supercontinent dating back to 300 million years ago. While this might at first seem rather innocuous information, the implications of starting a discussion about diversity with such a notion are quite significant. Visualizing the interlocked continents provides a basis for discussing the role of geography in determining the genetic variability in humans. CPSP would illuminate that the primordial, postmodern world was one without borders, both in the local and global sense. This suggests that as humanity developed, it was through various emergent social and cultural milieus that borders and "ownership" were developed. More specifically, maps and animations of Pangaea demonstrate two important aspects of human diversity. First, it emphasizes the interconnectivity of the world solely as it exists in nature, without colonialism and human influence. Second, discussing Pangaea has the potential of positioning human difference, such as skin color, hair, or facial features, not as a precondition for discrimination, but one of evolutionary survival. Thus, this information would be helpful not only in providing context for human diversity but also for the postmodern world of globalization and hybridity in which we now live.

Media Resources

Media resources are another way to stimulate discussions about diversity using CPSP. Videos such as *Race: The Power of an Illusion* produced by PBS, demonstrates a DNA lab activity that could be used in conjunction with the concepts of the above discussion. One aspect of the lab involves students finding out that there is actually more genetic variability *within* racial groups than there is *between* racial groups. This upends students' misconceptions about human differences that are based simply on appearance, as White students would likely find that their DNA sequence is more comparable to Black students than to other White students. According to the AAA's "Statement on 'Race'" (AAA, 1998), this is because the majority of genetic variation exists within racial categories:

> With the vast expansion of scientific knowledge in this century, however, it has become clear that human populations are not unambiguous, clearly demarcated, biologically distinct groups .... This means that there is greater variation within "racial" groups than between them.[10]

The AAA statement clearly indicates that throughout the history of humanity, the continued sharing of a common DNA has maintained all of humankind as a single species. Relatedly, another invaluable media source is *The Journey of Man: A Genetic Odyssey,* an award-winning book and documentary by Spencer Wells, a leading population geneticist and director of the Genographic Project from National Geographic. In this research, Wells travels around the world analyzing DNA of hundreds of thousands of participants in a quest to map the history of human migration (Maltby, 2003; Wells, 2003). All of these resources provide much-needed context for science educators to understand both themselves and their present-day classrooms.

The point here is that while these subtleties help explain human differences, which is important for understanding epidemiological trends, the same scientific knowledge can be used to support a more equitable and democratic human community. It is important to acknowledge the widespread implications of the distortions and omissions of (scientific) knowledge that has contributed to, as well as acquired from, the human experience that we all share.

Summary

This chapter explored the indistinguishable role of science for deconstructing narratives on race and human difference, both being contributors to systemic imparity in schools and society. Despite increasing discussions about student diversity in educational discourse, science education research often lacks critical, multicultural, and culturally relevant strategies and practices. Additionally, discussions on the specific use of science for increasing diversity awareness, cultural relevance, and linking students to common (global) humanity, are observably limited. To address some of these gaps in educational discourse, especially in relation to teacher preparation programs, this chapter presented critical postmodern science pedagogy, or CPSP, as a critical, transdisciplinary method of teaching about diversity.

History reminds us that change is constant. However, without critical and ongoing investigations of the past, the social construct of race will remain a socially debilitating cultural phenomenon that indisputably interferes with global posterity. As such, the importance of science literacy is increasingly important for 21st-century students, as well as humankind at-large. To that end, to avoid becoming an "essentializing" narrative, a CPSP approach should also include critiques of the larger science domain and interrogate the role of scientific information throughout history. As such, when using CPSP, it is important to remove the lens of racial monoliths when exploring diverse cultural groups. As with all academic subjects, the role of contextualization and proper historical alignment are imperative when talking about race, gender, ethnicity, or access.

The overall premise of CPSP suggests that progress is only as valuable as the extent to which the beliefs and actions of an individual or group are examined from within. This sort of critique is inseparable from the critical nature of CPSP. Using science as a stimulus for teaching about diversity has the potential to disentangle many misconceptions about human diversity by systematically addressing fundamental aspects of our shared humanity. This, in turn, leads to critical questions about the social and historical discourse of diversity in education. The next chapter outlines the research methods used to explore CPSP and respond to the larger study of this book.

Notes

1. Kemet, var. KMT; kmt, is term derived from Pharaonic language of the Ancient Egyptians, and was used to indicate blackness of the earth and the people of Africa (Diop, 1981, pp. 41–43).
2. Note: Diop's PhD dissertation was so controversial that it took several years and the support of scholars from a variety of fields to convince the institution of the validity of his work.
3. Mannheim (1954) suggests that investigation history (or, to Foucault, the effects of power) may be the closest one could come to taking an objective view (i.e., dispassionate; transcendental) of data.
4. "The brains of the different EUROPEAN NATIONS differ considerably from each other, but a common type characterizes them all, and distinguishes them from those now described. They are decidedly larger than

the Hindoo, American Indian, and Negro heads; and this indicates superior force of mental character. The portion before the ear, connected with the intellectual faculties, and the coronal region, or the organs of the moral sentiments, are more amply developed in proportion to the base and posterior inferior parts of the brain, the organs of the animal propensities. In short, they indicate a higher natural power of reflection, and a greater natural tendency to justice, benevolence, veneration, and refinement, than the others. The organs in which the European brain in an especial degree excels, are, Ideality, Conscientiousness, Causality, and Wit. The organs of these faculties are almost invariably small in barbarous and savage tribes. The European skull belongs to the Caucasian variety of Blumenbach, which he considers as the most beautiful and perfect of all the national crania in the world; and in this point he and the phrenologists agree" (Combe, 1830, p. 618).

5. The term eugenics was coined by Darwin's cousin, Francis Galton (Proctor, 1988).
6. Especially considering that this country has allocated nearly $705.4 billion for defense spending for the 2021 fiscal year (U.S. Department of Defense, 2020).
7. For Popper (1966), pseudo-rationalism is "the immodest belief in one's superior intellectual gifts, the claim to be initiated, to know with certainty, and with authority" adopted by Plato and others (p. 433).
8. Using the terms class and poverty as a euphemism for race, for example.
9. Early antiracists adopted a rigid stance that generally favored a Black-White binary view as the principal form critique of racism over cultural pluralism. Thus, having become trapped within a reductive, dichotomous metanarrative of racism that ignored postmodern identities (Hall, 1992) and the intersectionality of race, gender, and class oppressions (Collins, 2000) embedded within educational discourse, a solely antiracist approach to racial inequality would not suffice in a pluralistic society (May, 1999).
10. See also: AAPA Statement on Biological Aspects of Race (1996).

4 Research Context

This chapter provides a summary of the research method used in this study, which draws on the critical postmodern science framework of diversity presented in the previous chapter. This provides a foundational basis for the book's larger discussion. First, the study's purpose and guiding questions are reviewed, followed by a brief overview of qualitative research and case study research from a critical postmodern perspective, including the rationale of using this particular design. A description of the research design is also provided, including a detailed explanation of case and participant selection, data collection and data analysis procedures, strategies to ensure quality and trustworthiness, as well as book limitations and ethical considerations. The aim of this research study is outlined below is to address the following questions about science, diversity, and education:

1 *What is the role of science in teaching about diversity to undergraduate preservice teachers (at an urban research university in the Southeastern United States)?*
2 *How might this approach impact preservice teachers across disciplines in terms of raising greater critical awareness and influencing inclusive classroom practices?*

As mentioned in Chapter 1, the research questions were informed by the understanding that current demographic shifts among students do not culturally match the teaching force and a theoretical framework that recognizes the need for a critical, interpretive, and incisive approach to discussing diversity issues such as race and ethnicity (Stake, 1995). Along with the theoretical framework, the research questions helped shape the chosen unit of analysis through which the phenomenon of the utility of science as a critical method for teaching about diversity to preservice teachers (PSTs) were investigated. To thoroughly examine this potential role of science in what is here referred to as critical postmodern science pedagogy (CPSP), a single case study was conducted with three undergraduate courses designed for PSTs that address issues of diversity in education. To achieve this, a concerted examination of qualitative research and the method of inquiry is provided below.

Qualitative Research

Qualitative research is a systematic approach to understanding social phenomena through in-depth and detailed analyses of the human experience that can be understood as "a set of interpretive, material practices that make the world visible" (Denzin & Lincoln, 2005, p. 3). As a form of investigative inquiry, qualitative research seeks to understand a social or human problem through thick descriptions of phenomena as they occur in their natural settings (Creswell, 2009). Such investigations often involve a collective interpretation of meaning(s) constructed from the perspectives of both the qualitative researcher as well as participants (Gall, Gall, & Borg, 2005). Given the in-depth nature of using multiple data sources, an interpretive, qualitative method was used to address the research questions through detailed examinations of participants regarding the utility of science for teaching about diversity (Lincoln & Guba, 1985).

In qualitative research, there is an implication is that theory should be withheld during the investigative process to mitigate external influences on the outcome, as if one could approach such a task having no preconceptions (Moustakas, 1994). However, operating under the guise of value-neutrality, much as traditional versions of science is wont to do, it is not infrequent that qualitative research is clumsily burdened by dogged pursuits of unachievable transcendence or phenomenological abstraction (Atwater, Russell, & Butler, 2013; Carter, Larke, Singleton-Taylor, & Santos, 2003). This is not to say that such pursuits have no merit or that they cannot produce meaningful results. Qualitative investigations, such as ethnography, that honor the *emic* perspective of participants' lived experiences, as well as many empirical experiments or examinations, have contributed much to how we understand the social and natural worlds (Creswell, 2009; Wolcott, 1999).

Role of the Researchers

Qualitative research is unique in its ability to bring the researcher into a study, and with a first-account perspective that provides enhanced understandings of complex social phenomena (Creswell, 2007; Glesne, 2011; Marshall & Rossman, 2006; Patton, 1990). Since the researchers' subjective views and theoretical positioning are inextricably embedded within qualitative investigations, it is important to restate our proximity to the research. As mentioned in Chapter 1, we [the researchers] taught science, technology, and/or history courses in K-12 education for several years prior to this study. We taught diverse age groups, including elementary, middle, secondary, undergraduate, and graduate courses, all within the southeastern United States. Some of the undergraduate, graduate, and doctoral courses were at the proposed research site.

While having a strong familiarity with the issues and impacts of racial discrimination and exclusion in education was important, we also acknowledge the inexorable influence of bias on the research process. As such, we made attempts to bracket, or systematically account for, potentially biased views in the research process (Creswell, 2007; Kvale, 2007; Peshkin, 1988; Rubin & Rubin, 2005). Bracketing refers to the process of removing the researcher's subjectivities while preserving participants' interpretations and experiences with diversity issues in education (Moustakas, 1994). Although we cannot as composite social beings be fully abstracted from the human experience, acknowledging influences linked to the role of the researcher can help mitigate subjectivities brought to the process. There is an indissoluble link between researcher and research within the context of this study on science, diversity, and education. As such, the researchers' experiences are often doubly marginalized [Black-White biracial man, Black-Caribbean man, and Black woman], which is important to note.

This research stemmed from a perceptible desire to affect positive change as critical actors in education. The apparent prevalence of uncritical dispositions about diversity was of particular interest in this study. To presume for example, that one could perform any measure of inductive or deductive analysis that would produce, say, a grounded theory or theory of special relativity without having at least an uninflected tenor of bias imposed on the analyses seems a futile argument (Einstein, 1920; Glaser & Strauss, 1967). In other words, while it is important to take into account the effects of external influences during the investigative process and in the analysis of data, to suggest that theory or Self should (or could) be suspended from, rather than made explicit within and iteratively scrutinized for new directions and possible misconceptions, indeed, seems regressive. Thus, despite implications of value-laden views and inherent biases, critical and contextualized appraisals and a reflexive empathy for participants provided a meaningful basis for richer analyses. Furthermore, member-checking and other forms of triangulation helped account for any inherent bias and ensure the study was trustworthy and of high quality.

Criticalist Research

This section briefly outlines the epistemological basis for the book, which connects the previously discussed theoretical framework with the criticalist research design alluded to above. Criticalist research assumes socially and historically embedded disparities of power bound to ideologies and discourses that both construct and limit the subjugated individual through multiple and interconnected forms of oppression enacted by institutions, systems, and structures throughout society (Canella, 2007). The goal of this research was to develop incisive explanations regarding the unit of analysis that would lend insight beyond surface-level data from observations and participants' responses. This means examining the data as well as human interactions for latent oppressions and power differentials often subtly muted by social, cultural, or structural influences. Not only would the absence of such awareness work against an interpretive research design, but it also dispelled the opportunity to affect positive change in education as it is counterintuitive to the ancillary goal to help prepare PSTs to be critical stewards of posterity (Apple, 2010).

The CPSP framework outlined in the previous chapter ultimately directed this criticalist research paradigm, within which a single case study was used to explore the utility of science as a critical method to teach about diversity in undergraduate teacher education programs (TEPs). This position contests pathological forms of rationalism, objectivism, and logical positivism on the basis that there exists no singular objective truth and reality, but rather a series of social and cultural derivatives, interactions and institutions that (mis)shape our multiple perceptions of reality to situate the human subject within it (Atwater, 1996b; Haraway, 1988; Hartsock, 1989; Hutchison, 2006; Kincheloe, 2005; Lather, 1992; Ültanir, 2012). This suggests that any method of constructing, acquiring, sorting, analyzing, or transmitting knowledge is not a passive activity, and that social institutions necessarily influence and often attempt to direct social activity (Kincheloe, 2005; von Glasersfeld, 1984). Thus, from a CPSP perspective, within the irreducible plurality of contextual, often overlapping conceptual schemas re/constructed and shared among PSTs about race and ethnicity rests many "truths" and social realities that are perhaps more aptly understood as co-mediated inter/actions between the subject and the world, as well as the oppressions enacted by and/or upon them (Bernstein, 1983; Freire, 1970; Kincheloe, 2005).

An important implication posited by a CPSP framework is that research, instruction, and the production of knowledge are irrevocably intertwined in respective struggles against dominant class social and cultural reproduction in education, as well as the instruments, grand narratives, and ideological dimensions of power by which they are governed (Apple, 2000; Aronowitz & Giroux, 1991; Bourdieu & Passeron, 1990; Foucault, 1980, 2002; Freire, 1970; Kincheloe, 2005; Lyotard, 1984). As previously discussed, misapplications and wrongful positioning of scientific knowledge have traditionally privileged males of direct European descent, which often led to social distortion and stratification especially along the lines of race, gender, and sexuality (Campbell, Denes, & Morrison, 2000; Collins, 1998; Foeman, 2009; Larson, 1995;

Marks, 2008; Ordover, 2003; Witzig, 1996). This critical perspective suggests that the systematic process of scientific inquiry does not occur mutually exclusive of external factors. In contrast, when viewed as one of many social narratives, critical applications of scientific knowledge can help explain various causal links in the natural world, including the history and trajectory of humanity (AAAS, 2010; Aronowitz, 1988; Diop, 1974). However, much of the research that addresses diversity issues in teacher education does little to deconstruct social and historical elements of race and culture, particularly through a scientific lens. Guided by a CPSP framework, this case study sought to address some of these conceptual gaps, and thus, to contribute both practical and theoretical knowledge about the utility of science in grounding teacher practice in more critical and inclusive understandings of student diversity.

This task was accomplished through the use of descriptive and explanatory narratives derived from field notes, observations, interviews, and class discussions with three undergraduate teacher education courses. This approach allowed collective meaning(s) to be re/constructed in ways that respond to the role of science as a critical method for teaching about diversity and that inform the practice of PSTs as they prepare to enter diverse classrooms. Investigating the commonality among all humans with PSTs evoked critical questions and responses during the interviews and class discussions about social and political influences that have shaped history and continually shape the present. As suggested in the previous chapter, a CPSP perspective implies that such reflections would include critiques of the domain of science itself with respect to power and political economy, or the many injudicious uses of science that have historically undermined democratic race relations (Barton & McLaren, 2001; Marks, 1996; Smedley & Smedley, 2005). Furthermore, the propensity in U.S. society to either ignore, obscure, or reify issues of racism and other such hierarchical discriminations due to largely unexamined structural forces of the past suggests that examining the disproportionate effects of power within the context of science, education, and diversity is a pivotal component in PSTs' development of critical and robust understandings of race and ethnicity. Given the current lack of discourse and educational research in this regard, this book's findings suggest that the phenomenon of using science to teach about, deconstruct and discuss such issues from a critical postmodern perspective is a potentially viable option for achieving this goal, and in supporting human responsive, highly effective educators for the 21st century.

Broadly speaking, PSTs' social and cultural influences prior to pursuing a career in education offer important contextual clues about how and what views were formed about race and ethnicity, as well as their preparedness to teach in diverse classrooms (Garmon, 2004, 2005). Just as science has traditionally been subjected to whims of power exercised through dominant social institutions and cultural milieus, PSTs' views about race and ethnicity, either directly or indirectly, respond to the confluence of socially and historically contingent forces. Yet, in the social apertures formed by confronting scientific knowledge and teacher education as contested sites of knowledge production for pedagogical practice, context-specific realities are then re/formed, re/fashioned, and expanded relative to PSTs' particular worldviews and provide insight for preparing PSTs to teach in diverse settings.

Much in the way that multicultural science education endeavors to reconstitute misconceptions about the natural world by generating new scientific knowledge that incorporates the myriad cultural influences brought into the classroom, the impetus for this book was to offer new ways of initiating discussions about diversity in TEPs from a scientific perspective (Atwater, 1996b). To this end, the study sought to reconstruct participants' understanding of the social world, while also acknowledging social and historical contingencies that influence the production of knowledge (Denzin & Lincoln, 2005; Guba & Lincoln, 1994). The epistemological grounding of the study and its alignment with the theoretical framework allowed for a critical, interpretive exploration and description of meanings co- and re-constructed throughout the investigation from the perspective of both researcher and participants.

It is important to note that this book sought neither to constrain nor to reduce what is known about teacher education and human diversity down to strict generalizable principles, as positivist or purely universalist conceptualizations of science might otherwise be inclined to do (Loving, 1997; Matthews, 2008). Rather, it sought to explore the utility of science as a critical method to expand how complex features of knowledge production about diversity are discussed and confronted within the context of a multiply constructed, culturally nuanced narrative of humanity (Kincheloe, 2005). A central goal of the study was to demonstrate that by using science to assist in developing a more critical consciousness in teacher education about race and ethnicity, it is possible to co-construct new perspectives of what it means to teach in diverse settings in ways that are more democratic and inclusive. Through a CPSP lens, the present research suggests contextualized discussions, interactions and collaboration with PST participants while also attends to larger social and historical influences that negatively impact their views about diversity, and subsequently, their classroom practices in diverse settings (Garmon, 2004; Howard, 2006; Sleeter, 2001). That is, valuing participants' individual perspectives and descriptions encouraged critical reflections that provide deep and meaningful insights into their views of science, diversity, and education (Baxter & Jack, 2008; Lather, 1992). Therefore, this research is intended to contribute knowledge toward the positive transformation of teacher practice through a critical awareness and reflexivity about race and ethnicity afforded by critical applications of science in TEPs. Indeed, the contextual adaptability and depth of perspective afforded by an interpretive case study design presented such an opportunity (Stake, 1995, 2005; Yin, 2003).

Case Study Research

Case study research is a form of empirical qualitative inquiry used to address questions of complex social phenomena within a bounded system (Creswell, 1998; Yin, 2003). Framed within the particular time, space, and place in which they occur, case studies contribute to knowledge of individual, group, organizational, social, and political issues through multifaceted investigations of phenomena within authentic (naturalistic) contexts (Baxter & Jack, 2008; Miles & Huberman, 1994; Rowley, 2002, p. 16; Yin, 2003). Regarding their scope, Yin (2003) notes that case studies make deliberate use of these contextual conditions for their salience in understanding the phenomenon of the study and role in defining the often unclear boundaries between them. This, he argues, is because the "phenomenon and context are not always distinguishable in real-life situations," which necessitates in-depth investigations as a means to delineate between them (Yin, 2003, p. 13). As such, case study research is useful for exploring questions that seek to describe, explain, or evaluate an intervention, such as the understudied phenomenon of CPSP, as well as the context of the case and the site where the investigation occurred (Hartley, 2004; Yin, 2003).

Dobson (1999) offers that when conducting an interpretive case study, one should first state whether a descriptive or an explanatory approach is being used, where "description suggests the adoption of an idealist philosophy whilst explanation suggests a realist approach" (p. 268). However, as Luck, Jackson, and Usher (2006) suggest, the dynamic scope of case studies points to the dialectical interplay of structure and agency in interpretive educational research, which Broadfoot (2002, p. 5) contends is extant "between the external directives to institutions which shape the social space and the individual's capacity to choose; to be self-determining." One implication is that in-depth case studies are those in which the macro/collectivist perspective of structure and the micro/individualist perspective of agency are both taken into account during the research process (Archer, 1995; Bandura, 2000). By this measure, the critical postmodern perspective used in this study suggests that both description and explanation were necessary. On one hand, there is a need to explain macrostructural effects regarding the TEP itself, such as curriculum and supervised experiences. Yet, with respect to an equally vital need to respond to the individual agency and worldviews of PSTs, providing contextualized descriptions of these structural conditions would be more informative than structural explanations alone (Dobson, 1999; Kilduff & Mehra, 1997).

To capture both structural and agentic qualities[1] regarding PSTs, TEPs, and the understudied phenomenon of CPSP explored in this study, descriptive as well as explanatory elements of case study research were addressed. Whereas descriptive questions address the "what" of social phenomena, explanatory questions might examine the "how" or "why" of these phenomena (Boeije, 2010, p. 25). As mentioned in this chapter, the case for this research was framed within the context of one TEP in the College of Education at an urban research university in the Southeastern United States. This context defined the boundaries of the phenomenon under investigation – the role of science as a critical method for teaching about diversity. Three PST education courses that addressed social issues in education, such as student diversity, served as the unit of analysis through which the phenomenon was explored. Additionally, another point of inquiry considered how the use of science could potentially elucidate issues of race and ethnicity, which could assist in preparing PSTs to teach in diverse classroom settings. Overall, the study was interested in the efficacy of science in teaching PSTs about diversity, and how doing so would inform and promote more positive and inclusive teaching practices.

As noted, while it cannot directly answer questions of a personal or cultural nature, science can provide likely causal relationships from which to draw conclusions about ethical issues (AAAS, 2010). Both the descriptive and explanatory components of this exploratory case study assisted in linking the collected data with participants' perceptions about using science as a critical method to discuss diversity issues. The first research question is descriptive and seeks to glean rich interpretations about the role of science in teaching about diversity. Additionally, the first research question is rather exploratory, as it also proposes further inquiry along these lines (Yin, 2003, 2012). The second question is explanatory to the extent that the study seeks to predictively explain how and why using science to teach about diversity would be beneficial to PSTs, and general programs of education, in the future. In totality, this study served as a means to describe and explain the phenomenon of using science to discuss diversity issues in teacher education, and thus to contribute knowledge about moving 21st-century teacher education toward progressive, transformative, and inclusive teaching practices.

Case Study Design

Case studies are ideal for exploring unique or understudied phenomena (Hartley, 2004; Yin, 2003). Case study design is directed by the research questions and related theoretical perspectives, as well as the overall purpose of the investigation (Baxter & Jack, 2008; Yin, 2012). The research questions and purpose of this book were not exclusively concerned with the case of PST courses, but rather the phenomenon of CPSP in the PSTs classroom illustrated by the case. As such, the focus of the inquiry extended further than the case itself given that using science as a critical method to teach about diversity is neither indefinitely bound to the context, nor the unit of analysis. The case study design chosen to explore this topic was influenced by this need to examine and understand a phenomenon beyond the case.

This aligns with Stake's (1995) view that case studies can also be used as an instrument to understand phenomena beyond the case. Like most case studies, an instrumental case study explores a particular case within a bounded context to gain a better understanding of some phenomenon; though, rather than studying the case itself, the researcher is primarily interested in particular phenomena that can be accessed through the case under investigation (Stake, 2005). This book was interested in the phenomenon of science as a critical method to teach about diversity, or CPSP. An instrumental case study design allowed the researcher to focus on and gain insights about this specific external interest through a particular case (Luck et al., 2006). With this approach, it was not necessary to find a typical case by which to examine CPSP, but rather, one that might assist in providing insight into a related issue. Thus, while important for the implications CPSP may hold for TEPs in general, the case of three PST courses became secondary to the issue being examined and served as support rather than the focus (Stake, 2005). As a proxy to access the investigated phenomenon of CPSP, the unit of analysis supported an instrumental case study design.

Research Design

An effective research design involves creating a plan for establishing a logical connection between the research questions and the methods used to investigate and acquire supporting evidence from which to later draw conclusions (Yin, 2003). Such a plan for how a research study is conducted generally includes explicating the procedures for the collection, analysis and interpretation of data. According to Yin (2003), an effective case study design is comprised of five components: (a) the study's questions; (b) its propositions (if any); (c) its unit of analysis; (d) data-to-proposition linking logic; and (e) criteria for interpreting findings. Using an in-depth, single case design this study explored the role of using science to teach about diversity to PSTs and how doing so might inform their practice in positive and inclusive ways. The study was conducted in four phases (Appendix D) with three undergraduate teacher education courses that address student diversity.[2] Details of the research site, sampling strategy, and phases of data collection and data analysis are provided in the following sections.

Research Setting

To more closely examine the central phenomenon of science as a critical method for teaching about diversity, this case study was delimited to the College of Education at University of Tesaw [UT] – (pseudonym), an urban research university located in Wasetville (WSV) – (pseudonym), a large metropolitan area in the Southeastern United States. The university is located in a diverse city and is surrounded by many diverse urban communities. Many of the courses required for admission into the TEP and for teacher licensure address various aspects of student diversity, including race, class, gender, linguistics, and ability within the context of the 21st-century classroom. Likewise, reflection, instrumentation, and context are also important aspects of case study research (Yin, 2012).

In each of the courses, interactions between the researcher, students, and professors, as well as classroom discussions related to issues of race and ethnicity were critically and reflexively examined using instruments and methods designed to explore the larger issue regarding the efficacy of CPSP for addressing diversity issues in teacher education (Baxter & Jack, 2008). These instruments of inquiry included questionnaires, participant observations, interviews, and researcher-led class discussions. Each of these instruments provided valuable data with which to explore the central phenomenon of this study. By narrowing the focus of the investigation to a single case comprised of multiple data sources and overlapping aspects to the PST unit of analysis, the study was situated to yield thick, contextualized descriptions of the utility of CPSP in preparing teachers for diverse classroom settings (Baxter & Jack, 2008; Yin, 2003). This was made possible via critical, yet responsive observations, interactions, and discussions.

Context is another important aspect of qualitative research and case studies in particular. Case study contexts or backgrounds may include the social, historical, cultural, political, economic, ethical, aesthetic, and physical (Stake, 2005, p. 449). Several contextual features supported the selection of UT's College of Education as the primary research site in this study, including: the number of teachers produced in the state by UT, the student populations they are likely to encounter, and the demographic trends of PSTs in the TEP at UT. With respect to the study's focus on the role of science in teaching about diversity issues in teacher education, the sampling frame of PST courses was relevant to understanding this phenomenon for several reasons.

First, despite the evident ability of science to improve quality of life and the growing emphasis on producing a scientifically literate workforce, little is known about the utility of science for teaching about diversity. Second, as previously discussed, the distinct demographic homogeneity among U.S. teachers (i.e., predominately White females), coupled with the general lack of critical and contextual examinations of race and ethnicity in many TEPs, threatens the achievement in a rising number of classrooms with students from diverse cultural backgrounds (Howard, 2006; King, 1991; Milner, 2006; Swartz, 2003). Yet, pejorative misconceptions about the ability of students from racial and ethnic backgrounds different than their own continue to go largely unexamined in TEPs (Cochran-Smith & Zeichner, 2009; Howard, 2006). This apparent lack of critical self-examination in teacher education generally stems from and reinforces an all-but-imperceptible, internalized sense of privilege espoused by traditionally dominant White culture (King, 1991; McIntosh, 1989). The disproportionate number of cultural mismatches between students and teachers in diverse settings often comes at the expense of the agency, empowerment, and achievement of minority students. The expansion of science in society, growing classroom diversity, and documented incongruity of U.S. teachers and students provided significant insights about the role of science in teaching and learning in diverse settings.

As alluded in Chapter 1, education in contemporary society necessitates responsive educators grounded in critical thinking and rich, value-added democratic practices that honor diversity and social justice in the interest of all learners. The UT College of Education is nationally lauded for its high-quality TEP and progressive conceptual framework of professional, content, and pedagogical knowledge; demonstrated effectiveness through praxis and cultural competence; and strong commitment to making positive impacts in the classroom and the community. All of which suggests that by the end of their college careers, UT teacher graduates will be well-rounded practitioners. Thus, courses were also selected because one or more are required either to be admitted into or to obtain licensure from this TEP accredited by the *National Council for Accreditation of Teacher Education* (NCATE, 2011).

According to a recent letter from the president of the accrediting body sent to the institution,[3] the "accreditation decision indicates that the unit and its programs meet rigorous standards set forth by the professional education community" (Cibulka, 2014).[4] The accreditation serves as a form of insurance that the university's TEP has met rigorous national standards, as verified by an independent review by educational professionals, policymakers, and public stakeholders. The University of Tesaw assures that NCATE candidates are thoroughly assessed during their tenure and graduates from the TEP are well-prepared and thus highly sought after in the market. In addition to being nationally accredited at both the initial teacher preparation and advanced preparation levels, UT is in a state that leads the nation in *National Board Certified Teachers* (NBCT) and the institution itself is a leading provider of teachers to public school districts throughout the state (North Carolina Department of Public Instruction, 2019). Being located in a diverse urban city, many of the local schools to which PSTs are assigned for field experiences or student teaching, and districts in which many will likely begin their teaching careers, are racially and ethnically diverse urban schools. Along with the abovementioned commendations, this suggests a quality TEP that produces graduates who have the potential to generate a significant positive impact on minority students across the state.

Yet, when juxtaposed with rising classroom diversity, the fact that the overwhelming majority of U.S. teachers are White females who often find it difficult to connect with non-White students appears to belie this potentially positive impact (Brown, 2007; Delpit, 2006; Howard, 2006; King, 1991; Milner, 2006). Interestingly, this national trend of teacher demographics is reflected among current PSTs at UT. The percentage of White teachers across the nation is approximately 82.7%; 76% of whom are female (National Center for Education Statistics, 2017). Current UT institutional data reports a total enrollment of 29,615 students (University Profile, 2019). At the time of data collection, 26,571 students were enrolled at UT, with roughly 3.43% (992 students) undergraduates in the College of Education. Additional data indicate that 82.85% of PSTs at UT were White (812) and that 76.31% were female (757). In line with this trend, approximately 82.3% of 2013 teacher graduates were White, and roughly 72.43% were female ([University of Tesaw] Office of Institutional Research Data, 2014). As a note, the most recent institutional demographic data report 71.5% White (645) student enrollment, with 71.7% (591) of those being female ([University of Tesaw] Office of Institutional Research Data, 2017). In total, only 9% (78) of the total student population in the College of Education were men, which is consistent with national data trends (National Center for Education Statistics, 2017).

The implication is that the University of Tesaw's PSTs and teacher education graduates may also find it difficult to connect with students from cultural backgrounds different than their own. This point is particularly applicable given that while these PSTs (during field experiences) and new teacher graduates (at new jobs) primarily serve diverse student populations, they may espouse similarly uncritical assumptions about minority students' academic ability or willingness to learn (Delpit, 2006; King, 1991; Kunjufu, 2002).

The majority of these misconceptions are based on predisposed views of race and ethnicity formed prior to – and implicitly reinforced through – their TEPs, as such issues tend not to be discussed from a critical perspective (King, 1991; Sleeter, 2001; Swartz, 2003). By contrast, when combined with critical self-appraisals consistently high-quality instruction, a teacher's ability connect with students from a variety of cultural backgrounds not only supports democratic principles of self-efficacy and agency but can have demonstrable effects in the way of high academic achievement in high needs areas such as science (Chenoweth,

2007, 2009; Gay, 2002; Ladson-Billings, 1995a, 2009; Lee, 2001; Lee & Buxton, 2008). Rather than engender a strong and inclusive multiplier effect through the profession, however, the literature suggests that UT teacher graduates may instead hinder achievement, particularly in the state's more diverse school districts (Howard, 2006; King, 1991; Sleeter, 2001; Suriel & Atwater, 2012; Swartz, 2003).

The complexity of these issues suggested that a case study of PSTs at the University of Tesaw would produce valuable information about the potential role of science for teaching about race and ethnicity in TEPs, and also illuminate points of entry for CPSP beyond the case. Stake (2005) notes that instrumental case studies "draw the researcher toward illustrating how the concerns of researchers and theorists are manifest in the case" (p. 450). As such, this contextualization not only served to position this case study but by addressing a gap in the research and a critical need in education, reinforced its potential to contribute to an important knowledge base aimed at developing strategies to prepare PSTs for diverse settings. The following section details the purposive sampling strategy used to select the particular case, including the criteria for selecting participants.

Case Selection

Yin (2012) defines a case, or unit of analysis, as a bounded entity, such as "a person, organization, behavioral condition, event or other social phenomenon" (p. 6). Binding a case to a specific time, place, activity, or context centralizes the phenomenon under study, thereby limiting its scope and ultimately allowing for a more thorough investigation (Baxter & Jack, 2008). Given the essential role of boundaries in defining the case and determining what was studied, delimiting the study to explicit spatial and temporal dimensions was important for conducting a detailed investigation (Stake, 1995; Yin, 2003). The spatial context of the case consisted of three diversity-related courses designed for undergraduate PSTs in a single TEP at an urban research university. Temporally, the unit of analysis was bounded to the timeframe allotted for classroom observations, participant interviews and class discussions, which took place from November 2014 through January 2015 (Appendix D).

Furthermore, a key component of a qualitative sampling strategy is the extent to which it contributes to understanding the central phenomenon being investigated and responding to the research questions (Creswell, 2012). Of particular interest in this study was exploring the phenomenon of the potential role of science in teaching about diversity in the case of PST education courses. This sampling frame as well as the individual courses and participants that constitute the case in this study were purposefully selected for demographic and contextual representativeness, and overall relevance in developing an in-depth understanding of this phenomenon in the context of teacher education (Patton, 1990; Stake, 2005). By investigating the phenomenon of CPSP using multiple courses and participants but combined as a single case, the hope was that it would allow for thick descriptions of the potential role of science in reconstituting diversity discourse through critical reflection on the social and historical contexts of race and ethnicity (Baxter & Jack, 2008).

The role of CPSP was explored at the primary research site (UT's TEPs) with PSTs and professors in three undergraduate courses that address foundational issues in education aimed at establishing a baseline for their careers as professional educators. To bind the case and to assist with managing data collection and analysis processes, courses and participants included a convenience sample from a purposefully selected group of courses in the UT College of Education (Patton, 1990; Stake, 2005). A review of descriptions of undergraduate teacher education courses pointed to nine potential course candidates, from which three were chosen (Appendix E). Sampling criteria required that the courses were: (a) designed for undergraduate PSTs; (b) required either for admission into the TEP or for licensure; and (c) emphasized social and foundational issues in education such as student diversity and teaching in diverse settings.

The convenience sampling strategy was used to select the three different courses and participants consisted of solicitations and informed consent that included the purpose of the study, the research questions directing the study, as well as a brief explanation of the procedures that were used. Initial recruitment letters were sent via email to instructors of the different courses requesting: (a) to observe their class; (b) that they complete an online questionnaire, which included the informed consent notice; (c) that the researcher is allowed to conduct at least one class session; (d) to conduct an follow-up interview – included on the questionnaire, and (e) that they disseminate the questionnaire link to students. To ensure that the background of the study and the informed consent request was not overlooked, this information was also provided in the first section of the online questionnaire. To participate in the study, students had to meet specific criteria: (a) they had to be a student at UT, and enrolled in one of the courses selected by the researcher, and in which the instructor agreed to participate and (b) they had to be at least 18 years old. This section outlined case selection context and criteria. The following section details the data collection process used to explore the notion of CPSP in teacher education.

Data Collection

Creswell (2012) outlines five interrelated steps in the qualitative data collection process, which include: (1) identifying participants, research site, and sampling strategy; (2) obtaining permission to access the site and possible participants; (3) determining which data types are most salient for answering the research question; (4) designing protocols for data collection and recording; and (5) collecting data, with particular attention on ethical considerations (p. 205). Previously discussed were strategies used to identify the research site, the sampling frame for the PST courses, criteria for selecting study participants, and the method for soliciting and obtaining informed consent. This included explaining the contextual relevance of collecting data at the selected site and from this particular cohort of participants. This section discusses data collection procedures and rationale. These procedures were aimed at gathering rich case study data to respond to the research questions and central phenomenon of science as a critical method for teaching diversity in TEPs (Hancock & Algozzine, 2006).

To elicit rich descriptions that would assist in addressing this phenomenon, data were derived from multiple sources of evidence, including: participant and non-participant observations; an online questionnaire that captured participants' demographic information as well as brief narratives about diversity; a researcher-led class discussion that incorporated CPSP; and a semi-structured follow-up interview (Patton, 1990; Yin, 2003). Using multiple data sources to investigate an understudied phenomenon was a particular strength of case study research (Yin, 2003). Collected data included the researcher's field notes, audio recordings and transcripts, and participant questionnaire responses which were triangulated to increase the overall quality and reliability of the study (Yin, 2003). Triangulation of multiple sources of evidence helped ensure that collected data was accurate and aligned with the thrust of the inquiry through multiple measures of the same phenomenon (Yin, 2003, p. 99). In this case, field notes from observations were cross-referenced with questionnaires and transcribed interview data to determine the extent of convergence among participants' responses.

Questionnaires

After receiving confirmation from instructors who agreed to participate in the study, data collection commenced with classroom observations and a 10- to 20-minute online questionnaire. As mentioned, the questionnaire was designed to capture demographic data as well as assess participants' views regarding diversity-related issues such as race and ethnicity, and the potential role of science in discussing and teaching about them. Questionnaire responses also aided data triangulation with interview transcripts and field notes. In addition, the initial section of the questionnaires included background information about the study and an informed consent notice. The final section helped set up interview dates and times for those who agreed to one. The questionnaire was open-administered – meaning that participants completed them outside of class on their own time – during the period in which classroom observations were conducted. For those who agreed to be interviewed, questionnaire data were compared to interview responses to gauge any changes in participants' dispositions about science and diversity in teacher education. Responses were also used to triangulate observation and interview data. The researcher took all necessary precautions to ensure the confidentiality of the questionnaire data, by saving it on a secure web application platform accessed through a university account.

Observations

Questionnaires were completed within the first two to three weeks of the data collection process, during which time participant observations were also conducted. Observations took place in the UT College of Education. Each selected course was observed 2–4 times during their scheduled meeting times. Given that the purpose of this study was to explore the role of science for teaching about diversity, the focus of the observations was conceptualizing how and why this strategy might be important for PSTs for teaching in diverse settings, and what a sustained approach to doing so might look like. That is, what are the current perceptions of these PSTs about race and ethnicity, and how might infusing a critical scientific approach prove relevant for enhancing positive and inclusive teacher practices? Classes were observed from November 2014 to January 2015. Researcher reflections continued along this line of inquiry as questionnaire responses were submitted and the interviews began.

Interviews

Participating students and instructors were interviewed in person, by phone and electronically via Skype, and lasted an average of approximately 50 minutes. In-person interviews were conducted on campus, although a majority of the interviews were conducted either by phone or Skype. Participant responses were transcribed and personal identifiers were either changed or removed to maintain their anonymity. The semi-structured interview format assisted in gaining a deeper understanding of how participants view the role of science as a cross-curricular practice for teaching about diversity. This format also allowed more flexibility with interview protocol, in that it could borrow from questionnaire and observation data, as well as other interviews.

Instructional Intervention

At the discretion of each professor, the researcher led a class discussion using CPSP initially guided by the lesson plan in Appendix C. Each researcher-led session took place in the classroom of the respective course sections for a maximum duration of approximately 150 minutes per class. This was contingent on the instructors' course flexibility and the length of scheduled class meetings for each section. The researcher discussed beforehand with instructors their preferences for the class presentation and discussion so that it might blend into the current structure and pace of each course.

Throughout the data collection process, it was important to bear in mind that questions regarding race and ethnicity are sometimes sensitive in nature, and that students might be reluctant to discuss such topics either in class or in the interviews. It was anticipated that the timeframe would help as they were likely to be more comfortable by mid-semester. Thus, as part of the informed consent letter, participants were assured full confidentiality and were notified of the option to review any transcribed data. In addition, the researcher took pains to be responsive when addressing sensitive topics, while still adhering to the goal of in-depth data analyses that respond to the study's research questions.

Data Analysis

As earlier suggested, the primary strength of the data collection strategy was its multiplicity of methods. By drawing from multiple data measures to understand the singular phenomenon of CPSP in teacher education, this case study provided a basis for rigor and quality on an understudied topic. This laid the groundwork for rich analyses of the findings, and thus a greater understanding of the potential utility of this approach to diversity education purpose and driving inquiry of this study.

In qualitative research, the iterative process of data analysis begins at the onset of the study (Dey, 1993; Sandelowski, 1995; Stake, 1995). This includes general perceptions by the researcher regarding thematic connections that emerge during observations and interactions with participants. Such analyses, for Stake (1995), are "a matter of giving meaning to first impressions as well as to final compilations" (p. 71). During observations, for example, these first impressions stemmed from questions such as: (a) *What's going on in this situation, or that situation?* (b) *What does that behavior, phrase, body language, gesture mean?* and (c) *What symbols are used to illustrate meaning, and what are they?* (DeWalt & DeWalt, 2011; Spradley, 1980). In general, data were examined using Miles and Huberman's (1994) and Braun and Clarke's (2006) data analysis procedures similarly consisting of reduction, display, and conclusion drawing.

The initial open-coding process incorporated constant comparative analysis to simplify and focus the data from questionnaire responses, observation field notes, and interview transcripts into manageable segments and categories. Constant comparative analysis helps reduce data through the iterative process of coding and recoding (Glaser & Strauss, 1967). Although the constant comparative method is generally associated with the inductive development of grounded theories, its use is not restricted to such designs (Glaser & Strauss, 1967; O'Connor, Netting, & Thomas, 2008). The implication being the constant comparative method may have applications in naturalistic inquiry beyond the production of theory as a way of maintaining an emic perspective during qualitative investigations to allow for more in-depth analyses, for instance (Fram, 2013). Data analysis also included Spradley's (1980) semantic relationships of inclusion and attribution, in which coded data were arranged using Microsoft Excel (MSExcel) and qualitative software, according to segmented categories that corresponded with topics related to science and human diversity and in the context of teacher education (i.e., segment X is a [form of] Y). In this way, throughout the study, questionnaire responses, interview recordings and field notes were also coded and analyzed.

Data Analysis Procedures

To summarize and conceptualize collected data, Miles and Huberman's (1994) data analysis procedures of reduction, display, and conclusion were implemented. This involved anticipated findings and experiences relative to data collection, as well as how they were managed and interpreted afterward (DeWalt & DeWalt, 2011). Atlas.ti and MSExcel software were used for the coding and analysis, which involved data organization and textual analyses, as well as generating and organizing themes and topics relative to the research questions and purpose of this study. All data were securely stored in digital format on a password-protected computer. The following steps explain how understandings of the observed phenomena were developed during this phase of the study.

Reduction

Among the initial steps in the analysis process was data reduction, which generally involves simplifying and focusing data to make it more manageable (Miles & Huberman, 1994). This included, for example, elaborating on general and specific aspects of an experience and searching for commonalities and patterns in the data by making memos about recurring segments/categories and linking related quotations. The related process of frequency and declaration or enumeration (of quotes/phrases) was also used to detect specific patterns in the data (LeCompte, 2000). As part of the reduction process, data were analyzed from a CPSP perspective using constant comparative coding method which began with open and axial coding of observation and interview transcripts (Strauss, 1990; Strauss & Corbin, 1990). During the open coding process, data were "chunked," or divided, into several broad categories using related to science as well as issues of race and ethnicity in teacher education

(Glaser, 1978). Next, axial coding was applied, which is "a set of procedures whereby data are put back together in new ways after open coding, by making connections between categories" (Strauss & Corbin, 1990, p. 96). This allowed data to be arranged and categorized into observable patterns. Data chunks and codes were iteratively compared throughout this process, from which code segments were then formed into core categories, and the remaining categories into themes (Glaser, 1978; Glaser & Strauss, 1967; Spradley, 1980).

A preliminary step to this purposeful management of data involved a textual analysis of documentation searching for frequently occurring words, as well as in-context quotes and descriptions of situations. This produced a list of descriptive terms and phrases that were then indexed into categories developed from field notes, transcriptions, and reflections on these data from a critical postmodern science perspective (DeWalt & DeWalt, 2011). The goal was to develop a list of codes and definitions that could also be used as cover terms for particular domains, with respect to science, diversity, and education (Spradley, 1980). Next, using Spradley's (1980) semantic relationships of inclusion and attribution, data were re/arranged within and between code segments and categories, in addition to being checked against field notes and transcripts to ensure accuracy, and then re/coded against the indexed categories. For instance, where *code segment* **X** *is a type of* **Y** *relationship*, "**X**" might be a recurring phrase among PSTs or researcher observation regarding discussions of race and ethnicity in the classroom and/or when using CPSP; whereas "**Y**" might refer to an interconnecting conceptual or emotional response, such as an "identity reference" or "ambivalence toward issues of race." Resulting categories were grouped into taxonomic categories based on contextual relevance to the research questions as well as the frequency of their occurrence (LeCompte, 2000). These categories assisted in the formation of patterns/themes to help provide insights for understanding the potential role of science in the preparation of PSTs for diverse settings.

Displaying the Data

The process of reduction continued into the second step, where data were re/coded and displayed using Atlas.ti, MSExcel, and CMapTools to create semantic concept maps and tables. Doing so assisted in identifying patterns/themes through both visual and numerical representations (Daley, 2004; LeCompte, 2000). Visual displays of data were used throughout the research process, from initial conceptualization to data organization and analysis, which helped make deep conceptual connections beyond the text (Daley, 2004; Miles & Huberman, 1994). In this particular case, patterns were identified among categories that emerged from the coded segments, and those, from the raw data (Strauss & Corbin, 1990). Categories were then "collapsed" into larger components and re/arranged into preliminary themes related to the central phenomenon in this study of CPSP. Resulting taxonomies of semantic relationships were then illustrated as a concept map, table, or other such visual representation. The result of the coding and analysis process was the identification of themes that illuminate the potential role of CPSP in deconstructing diversity discourse in teacher education, and to offer in its place a reconstituted vision of human diversity grounded in critical, compassionate, and inclusive teaching practices.

Drawing Conclusions and Verification

Not only did visual displays help reduce and organize data, but they were also instrumental for interpretation and drawing conclusions throughout the analysis process (DeWalt & DeWalt, 2011; Miles & Huberman, 1994). Interpretations of the study findings were derived from these various analyses, which were ongoing and primarily involved comparing participant data/responses to explore the role and impact of using science to teach about diversity.

As the previous sections allude, the process of drawing and verifying conclusions in this study included numerical data. However, it is not wholly uncommon in qualitative research, and particularly case study designs, to incorporate various quantitative methods (Yin, 2003). While it is understood that numbers are merely representative and hold neither intrinsic nor existential meaning, not only did the supplementary enumeration make the data more manageable, but it also helped reveal unforeseen patterns that could be verified against qualitatively obtained data. As Miles and Huberman (1994) note, "qualitative analysis of all data with the aid of numbers is a good way of testing for possible bias, and seeing how robust our insights are" (p. 254). This in turn helped offset inherent biases in the research process thereby increasing the trustworthiness of the study.

Trustworthiness

Yin (2003) offers that "the development of case study designs needs to maximize four conditions related to design quality: (a) construct validity, (b) internal validity (for explanatory or causal case studies only), (c) external validity, and (d) reliability" (p. 19). To increase the validity of the findings and to strengthen the overall credibility of the study, data were triangulated using observation notes, interviews, participant responses and narratives, and member checking (Creswell & Miller, 2000; Seale, 1999; Stake, 2010). In addition, as the interview protocols were semi-structured, they allowed for questions to be derived from field notes and questionnaires. This provided opportunities for consulting the data to see if it aligned with participant responses.

In all, the researcher strove to reflectively consider the intended audience as well as questions of purpose, voice, and interpretation across field notes, interview transcriptions, and narratives (Creswell, 2012; Polkinghorne, 2003). Further, participants had the opportunity to discuss concluding analyses before being made available to the school, community, or for publication. The aim of which was to help ensure that, as human subjects, participants and their views remained central in this investigation while providing conceptual space for in-depth data analyses that responded to the research purpose and questions regarding the role of science in teaching about diversity.

Ethical Considerations

When conducting qualitative research with human subjects ethical considerations should be the highest priority (Yin, 2012). As the "primary vehicle for emic inquiry," this is especially true in case studies as they often involve sustained interactions with participants (Lincoln & Guba, 1985, p. 359). As such, the anonymity and confidentiality of participant data were a primary concern in this study. In addition, as earlier mentioned, it was important to consider that issues of race and ethnicity may be a sensitive area of discussion for some participants. Therefore, being responsive to this reality was also a goal throughout the study. Overall, this study imposed no perceivable risks to participants, nor did it pose any harm, although it may offer several benefits to TEPs and classroom instruction, particularly those in diverse settings.

At the outset of the study, participants were asked to provide informed consent, which was acquired prior to beginning the online questionnaire. Hard copy versions of the informed consent agreements were also available as an alternative to the online form. Participants could print the agreement from the online questionnaire. Information about the risks, benefits, and study design were detailed on this form. Participants were also provided with the researcher's contact information for any additional questions they might have had about the study. Only participants who were at least 18 years old were allowed to participate in the study; thus, parental consent was not required. In addition, they were free to withdraw from the study at any time.

With respect to data collection procedures, several precautions were taken to preserve participants' anonymity and confidentiality. For instance, the online form was located on a secure (remote) server using an account registered with the University of Tesaw, which meant that participants had to use their institutional login information to access the web application. Upon logging in to complete the form, participants' email addresses were captured to verify their affiliation with UT, which was required to complete the questionnaire. This also served as a means to create a unique identifier during data collection to keep participant information linked; however, at the time of data analysis, participants' email

addresses and other identification were de-identified. Questionnaire responses were accessed through the university's secure network and saved on a spreadsheet in a university provided Google Drive account, visible only to the researcher via a password-protected computer.

Furthermore, prior to all interviews, verbal consent was requested as interviews were audio recorded. Audio files and transcribed interviews were kept on a password protected computer that was only be accessed by the researcher. After being transcribed, audio recordings were deleted. Participants were also notified of the option to view transcriptions. To preserve participant anonymity, email addresses and pseudonyms were used in the data analysis and write-up rather than their real names. After questionnaires and interviews were completed and responses from both instruments matched, email addresses were deleted and replaced with the randomly generated number. Hard copies of any collected data were destroyed after three years by shredding. Data collected from participants may be used for professional development and publication purposes; however, participants were made aware of the potential use of the anonymized data for future research. After six years, all electronic files were deleted.

Potential Benefits

CPSP promotes developing a critical social consciousness about issues of race, ethnicity, and culture as they pertain to education. Thus, while the study posed no risk or harm to participants, there are many potential benefits of using science to teach about diversity in TEPs. The primary benefit of this research, both to participants and to society, is raising awareness about diversity issues in ways that promote positive and inclusive teacher practices. PSTs would directly benefit from such understandings, as it would in turn enhance their ability to reach students from diverse backgrounds. Similarly, this research offers a powerful additive to PST educators in the TEP at the University of Tesaw in preparing teachers to be successful in diverse classroom settings.

Benefits to society include the ability of more informed teachers and teacher educators to prepare a diverse and critically informed workforce and citizenry, which ultimately supports the cultivation of a democratic society. Used as a critical method in teacher education, science can increase PSTs' awareness and understanding of critical aspects of race and ethnicity underlying the more prevalent discourse surrounding diversity. It is anticipated that doing so would inform their practice in positive and inclusive ways that strengthen student–teacher relations and promote high academic performance in diverse settings, across subject areas. Ancillary benefits of this study include promoting critical-thinking, scientific literacy, and collaboration among, within and between classrooms and schools. Such benefits inherently support the nation's goal of growing a critical and pluralistic citizenry able to meet the challenges of a knowledge-based society centered on math and science skills.

Table 4.1 Student participant profiles (pseudonyms)

Demographics					School Data			Preferred Teaching		Diversity				
Name	Age	Sex	Race	F/R Lunch	HS Locale	Year	Course Section	Locale	Level	Lang. Spoken	Diverse HS Pop	Diverse HS Staff	Diverse Comm.	UT Racial Groups
Sara	30+	F	B	Y	R	JR	A	NP	HS	1	Y	Y	N	Many
Jess	20	F	W	Y	R	SO	A	R	HS	2	N	N	N	Prim. One
Allie	21	F	W	N	R	JR	A	NP	HE	1	N	N	N	Many
Matt	25	M	W	N	S	JR	B	NP	HS	1	N	N	N	Many
Kara	18	F	W	N	S	FR	B	R	HS	1	N	N	N	Many
Bertie	19	F	W	N	R	FR	B	U	HS	1	Y	Y	Y	Many
Silva	19	F	L/H	Y	S	SO	B	U	MS	2	Y	N	Y	Many
Alexi	22	F	W	Y	S	JR	C	S	HS/ HE	2	N	N	Y	Many
Ava	20+	F	W	N	S	JR	C	NP	HS	1	Y	Y	N	Many
Meg	20+	F	W	N	S	JR	C	S	MS/ HS	1	Y	Somewhat	Y	Two+

Context and Unit of Analysis

As stated in the previous chapter, the research setting for this study was an urban research university located in a diverse metropolitan area in the Southeastern United States. The unit analysis in this study consisted of ten (10) students and two (2) instructors from three diversity-related courses designed for undergraduate PSTs. The findings report analyses of questionnaire and interview data gathered from these twelve participants following a researcher-led class discussion regarding human diversity that used CPSP strategies. Table 4.1 provides a profile of each of the student participants, including their respective course sections. For anonymity, participants were given pseudonyms.

Student Descriptions

Sara

Sara is a mother of two from Selma, Alabama. She transferred to UT and is now a junior pursuing a degree in mathematics with an education minor. Sara comes from a strong, family oriented background, and describes herself as a very outgoing person who enjoys helping others. She and her mother moved a lot when she was younger and she attended various schools in the process.

Jess

Jess is from a small town in southwest Ohio. She's a sophomore who transferred to UT to pursue a degree in Spanish. She is very adamant about the positive role of education in one's life and discussed some of the perceived issues that impede this potential. Despite having little exposure to diverse settings growing up or in her college classes, Jess expresses an eagerness to support students to help them be the best that they can be, regardless of their background.

Allie

Allie is a junior majoring in English. She grew up in a small town in North Carolina which she describes as very religious. Allie experienced little in the way of diversity growing up and describes her cultural background as a "modern American family." Allie prefers to work with older students, and her goal is to become a college professor.

Matt

Matt is a junior majoring in history, who lived in Texas and Pennsylvania before moving to a city just outside of Wasetville (WSV). He describes his background as "bland, middle-class Anglo-Saxon." Matt is passionate about education and expresses a strong desire to be a history teacher that not only discusses the past but one concerned with reaching all students and helping them investigate and unpack historical "truths."

Kara

Kara is a first year student majoring in English, who grew up in a small town not far from WSV and attended Christian schools with little diversity through Grade 8. This lack of exposure to diversity changed when she started high school. And although the shift to a setting with more cultural variety was initially a shock, it seemed to be an overall positive experience.

Bertie

Bertie is a first year student majoring in history. She grew up in a small town near WSV which she describes as "not very diverse." She attended predominately White schools and describes her family and surroundings as "people the same race and ethnicity and background" as herself.

Silva

Silva is a sophomore majoring in Spanish. She is originally from Puerto Rico but moved to the United States when she was nine years old. Her experiences in with diversity in Puerto Rico were directed in large part by a lack of exposure and her Catholic upbringing.

Alexi

Alexi is a junior majoring in Chemistry. She was born in Ukraine, but spent most of her life in Florida "in a very traditional Russian home." Many of her early insights and experiences with diversity were influenced by a strong presence of Russian culture at home, as well as the magnet schools she attended in Florida. It wasn't until going to college that she was exposed to greater diversity.

Ava

Ava is a mother and a co-owner of a business with her husband. She decided to go back to school because she always wanted to teach and is now a junior majoring in English. Ava was "raised by two extremely Christian parents" and describes her upbringing as "very sheltered." She conveyed that her initial views of human diversity were shaped by her parents, whom she describes as "narrow-minded" about such issues. She also indicated that increased cultural exposure over time was ultimately more influential on her current views of diversity.

Meg

Meg is a junior majoring in French. She was raised in California and then lived in the Washington D.C. area for 20 years before moving to a city outside of WSV seven years ago. Meg's views on diversity were informed by her experiences with members of Latino/a, Asian, and gay communities while living in California. She describes her views about diversity as being firstly influenced by stereotypes learned from her family but eventually shifting after personal experiences.

Instructor Descriptions

Jonathan

Instructor Jonathan is an associate professor at UT, who specializes in science education. He is originally from Ghana, West Africa, but has been living in the United States for approximately 25 years. Jonathan is a molecular biologist, by trade, and has been involved in education in various capacities for approximately 45 years.

Tamela

Instructor Tamela is an adjunct education faculty member at UT. She was born and raised in Ohio and has been an educator for more than 20 years. Tamela has a master's degree in Curriculum and Instruction and has taught at every grade level, from K-12 to higher education. She is also certified in several subject areas, including: reading, middle grades English and mathematics and elementary education, and holds several other educational certifications (Table 4.2).

Table 4.2 Instructor participant profiles (pseudonyms)

Name (Pseudonym)	*Title*	*Years in Education*	*Subjects Taught*
Jonathan	Associate Professor	45+	Molecular Biology
Tamela	Instructor, Adjunct	20+	Middle Grades Education

Thematic Findings

This section presents the findings of the study. Five major themes emerged from the analysis of twelve (12) participant interviews regarding the utility of science in teaching about diversity, or CPSP (see Figure 4.1). Again, major themes that emerged from the analysis were: *Critical Challenges and Sociocultural Dissonance*; *"Big Picture" Understandings of Race, Ethnicity, and Culture; Racial Anxiety and Apprehension versus Meaningful Dialogue*; *Applications, Strategies, and Impacts.*

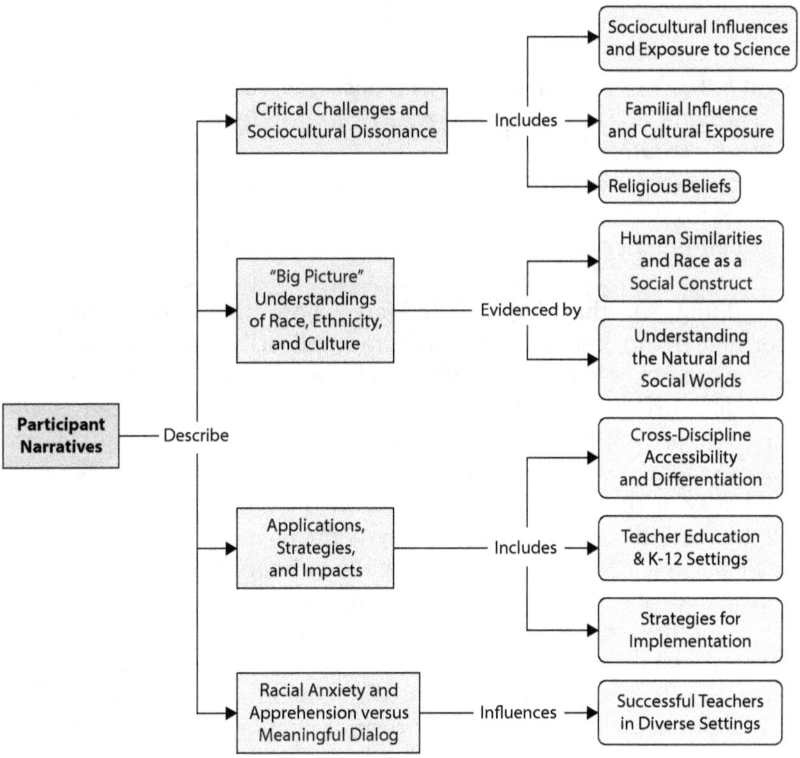

Figure 4.1 Semantic concept map of major themes and subthemes

These themes were interpreted as coming from both participants' and the researcher's perspectives regarding the role of science in teaching about diversity and the potential impact of this approach in raising critical awareness among PSTs and influencing positive and inclusive classroom practices (Patton, 1990; Rossman & Rallis, 2003).

Summary

Chapter 3 provided a summary of the research method used in this study, which draws on the critical postmodern science framework of diversity earlier presented. The research purpose and guiding questions for this book were reviewed, followed by a brief overview of qualitative research and case study research from a critical postmodern perspective, which included the rationale of using this particular design. Also provided was a description of the research design and procedures, including a detailed explanation of case and participant selection, data collection and data analysis procedures, strategies to ensure quality and trustworthiness, as well as study limitations, ethical considerations and potential benefits of this research. The primary focus of this book and the use of case study design was to address the following research questions:

1 *What is the role of science in teaching about diversity to undergraduate PSTs at an urban research university in the Southeastern United States?*
2 *How might this approach impact PSTs across disciplines in terms of raising greater critical awareness and influencing inclusive classroom practices?*

In doing so, the chapter illustrated how these questions were informed by an understanding of the implications of uncritical cultural mismatches between students and teachers, and a multiperspectival framework that recognizes the need for a critical, interpretive, and incisive approach to discussing diversity issues such as race and ethnicity (Kincheloe & Berry, 2004; Stake, 1995). Also discussed was the link between the theoretical framework, the research questions, and how through the chosen unit of analysis the phenomenon of the utility of science as a critical method for teaching about diversity to PSTs was investigated. By outlining and connecting the research design to this line of inquiry, this chapter provided the basis for an in-depth, qualitative case study that explores the role of CPSP (Yin, 2003). Chapter 5 discusses relevant findings, followed by Chapter 6 which synthesizes these findings, with respect to the overall purpose and research questions that guided the scope of the book.

Notes

1. Structure and agency conflicts/alignments in TEPs, with regard to teaching about diversity, etc.
2. Individual participants will include one professor per course (3 total), and up to five (5) students per course (15 max.). Data collection will include two (2) classroom observations, two (2) online questionnaires, two (2) semi-structured interviews, and at least one (1) instructional intervention in the form of a researcher-led class discussion – conducted with the entire class at the discretion of the professor.
3. Continuous Improvement Commission of the *Council for the Accreditation of Educator Preparation* (CAEP).
4. http://education.uncc.edu/sites/education.uncc.edu/files/media/Action%20Letter%20NCATE.pdf

5 Challenges and Understandings toward an Anti-Racist Education

The previous chapters discussed interrelated issues of science and diversity within the larger context of 21st-century schools and society, providing a basis for this book. The primary issue this book sought to address was the need for effective teachers in diverse settings to support a 21st-century workforce. Also provided in Chapter 1 was the purpose of the study, the research questions guiding the study, major terms used, the study's significance, as well as its delimitations and limitations. Chapters 2 and 3 explored through a critical postmodern lens the potential of science to deconstruct hegemonic narratives of race and human difference that have historically undermined racial parity in schools and society. In addition, Chapter 2 presented critical postmodern science pedagogy (CPSP) as a critical, transdisciplinary method of teaching about diversity.

Drawing from this framework, Chapter 4 outlined the research method used in this study. The research purpose and questions for the study were reviewed, followed by an overview of qualitative research and case study research, including the rationale for the selected approach. A description of the research design and procedures were also provided, as well as limitations and ethical considerations.

This chapter presents the findings from ten (10) semi-structured interviews conducted with undergraduate preservice teachers (PSTs) and two (2) instructors from three different diversity courses in the teacher education program (TEP) at the University of Tesaw (UT), as well as questionnaire responses and observation notes. Chapter 6 analyzes and discusses the findings and connects them to the following research questions:

1. *What is the role of science in teaching about diversity to undergraduate PSTs at an urban research university in the Southeastern United States?*
2. *How might this approach impact PSTs across disciplines in terms of raising greater critical awareness and influencing inclusive classroom practices?*

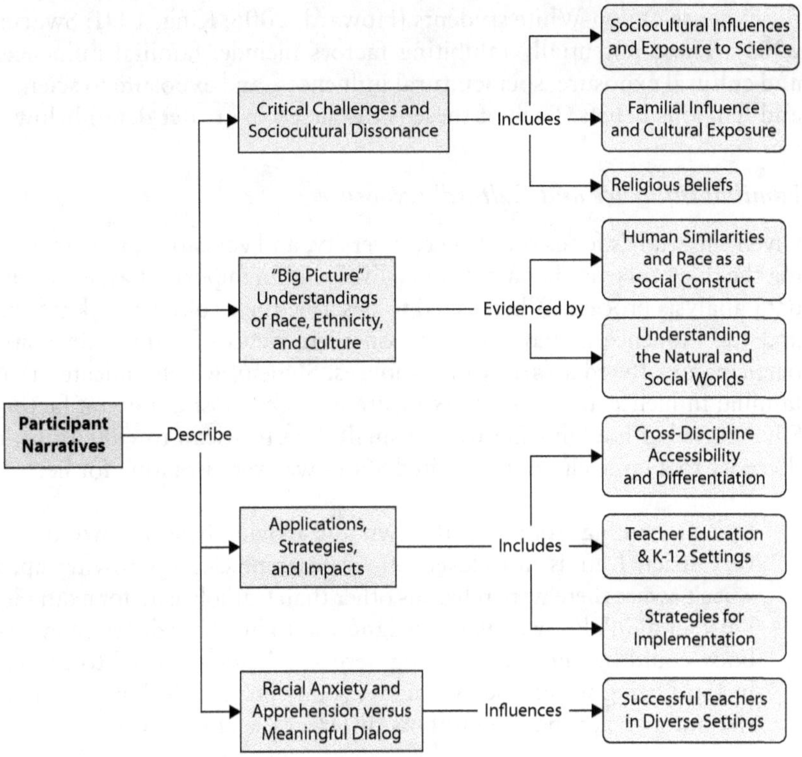

Figure 5.1 Semantic concept map of major themes and subthemes

This chapter discusses salient themes that emerged from analyses of the data. Four major themes emerged during the data analysis process. The major themes were: *Critical Challenges and Sociocultural Dissonance*; "*Big Picture*" *Understandings of Race, Ethnicity, and Culture*; *Racial Anxiety and Apprehension versus Meaningful Dialogue*; *Applications, Strategies, and Impacts*. Figure 5.1 (*restated from* Figure 4.1) illustrates the semantic relationships between participant narratives and major themes (green/rectangular). The findings suggest that science has the potential to contribute to education in meaningful ways.

Critical Challenges and Sociocultural Dissonance

This major theme sets the tone for the remainder of the findings. It emerged from participant narratives that describe critical challenges to understanding human diversity resulting from sociocultural dissonance, or difficulty reconciling unexpected or unexplained (or unexamined) cultural transitions, and potential effect on PSTs (Macdonald, 1998). When brought into the classroom, the negative effects of which often come at the expense of non-White students (Howard, 2006; King, 1991; Swartz, 2003). These potentially inhibiting factors include: familial influences and cultural exposure, sociocultural influences, and exposure to science, and religious beliefs. Each of these is discussed in greater detail below.

Familial Influence and Cultural Exposure

Given the study's focus on science, diversity, and education, understanding the role of science in participants' lives was an important aspect of the data analysis process. When asked to describe their cultural background and the influence it may have had on their views of human diversity, participants offered a variety of responses. Some of which indicated that familial influence and cultural exposure were the most dominant factors Silva describes how moving from a small, Puerto Rican town with little diversity to a large city in the United States was very isolating for her:

> I mean, coming from a small town and moving here, my world was very strict. I guess just closed off to different views. Growing up, I wasn't aware there were religions other than Catholicism, for example. I just – I didn't know. I was very ignorant to just how different everybody could be and then moving here and being exposed to things, but still being so secluded from everybody just for the language barrier and like getting used to how different the culture was.

Despite growing up in a diverse setting, Meg explains that her views of race and ethnicity were largely influenced by family members and cultural exposure:

> The community where I grew up, Orange County, pretty much developed as people who were fleeing the increasing diversity in Los Angeles. You've heard the term "White flight." So, Orange County was like a White flight place. Then, California was becoming more and more diverse. So, where I grew up, my high school, I wrote [on the questionnaire] that it was a pretty diverse place because there were tons of Latinos and tons of Asians. Like, all different walks of life, but not very many African-Americans – but tons of other types of diversity ... Oh, and the gays, actually. We had tons and tons of gay people.
>
> Everybody was economically pretty much the same actually, but there were not very many African-American people, because – I think because of the original White flight thing. We were very close to Compton. In fact, my grandparents owned an apartment building in Compton. I grew up thinking that I didn't have any racial bias against African-American people until I moved to D.C., and I realized that I had a fear when I saw African-American people. I

recognized it. It took me a while to recognize it, but I didn't realize I had it until I spent time with African-American people. Then I had to be like, "Okay, you're being stupid. This is ridiculous." There was an African-American family across the street from us. They were college professors at Cal. State Long Beach. They had more education than my whole family combined. I had to think that through, and then I was, "Okay, you're being ridiculous." This was a time of Marion Barry in D.C. and the whole crack pipe stuff. Initially, it was scary. Anyway, that's where I came from.

Researcher: What provoked that fear? Have you investigated that further?

Meg: I just felt like it was healthy to have a fear if you are in a dangerous situation, but it wasn't related to a person's skin color. It was more like, "You probably shouldn't be in Alexandria at 11 o'clock at night getting gas all by yourself. You should probably be in a better neighborhood that's well lit." I think it was from the news and just hearing about gangs and stuff on the news all the time. I didn't really realize – then, actually – I'm sure my parents were racists, but I just didn't know it growing up because we didn't have any Black people in our neighborhood. When we were in California for a funeral three years ago, everybody was just talking really racist. My daughter was with me, and we were like "Whoa!" People were saying things like, "I work with this one Black woman, and she always helps the Black families first, and I can tell she's always nicer to them." These were things that I just wouldn't have anticipated, and I realized at the time that I must have been raised with all of it. Oh, my gosh! I just remembered something, actually. I actually did find – my mom sent me a packet that had all my papers from when I was a kid. I had written a paper. I'm embarrassed to admit this …

I was a second grader, who really wasn't aware that there were such racists. I wrote this paper. I was in the gifted and talented classes, and we had this writing prompt. I remember it, actually. The writing prompt was "Which is quicker? Black or yellow?" That was the writing prompt. You just write whatever.

You don't have to really use your imagination to answer that. I can't even believe I'm saying this out loud, but I wrote "Which is quicker to ruin a neighborhood?"

Researcher: In second grade?

Meg: My daughter is laughing at me. Yes. I just remember at the time going back when I read this paper five or six years ago, I'm like, "Oh, my gosh!" My grandparents had owned that apartment building in Compton, and then they started building public housing in Compton. My grandparents sold their building because of what was happening. They must have been talking about what was happening to that neighborhood and stuff and blaming it on Black people. I was picking up on that, but I wrote an essay about the Black people ruining the neighborhood in Compton, and Chinese people – as a second-grader, I was identifying Chinese people as yellow people. Like, "They seem to be fine, upstanding citizens." I don't know. I don't remember the exact text.

Researcher: That's interesting. It also speaks to the fact that it is something we absorb, knowingly or not, from our environment. Sounds like an example of that.

Meg: It's a perfect example of that. Yeah, we have that, and we don't realize that we have that until you're called out on it – this fear based on ignorance. I got over it once I named it when I moved to D.C.

Meg adds that she's been rather surprised at the amount of racism still lingering in the South, and the assumption that, because she was also White, she would agree with racial epithets and slurs uttered by Whites from the South.

> Truth be told, I never heard anybody say the "N-word" out loud until I moved to [this state]. People just dropped it in the conversation. I literally cannot even say that word out loud. I can't even believe – It took me off guard. Then, about two years ago, my kids and I – and my husband – we just said, "We're not going to tolerate it." When people say really horrible things, we are going to say, 'It's horrible, and I don't agree with you.' Try not to make a huge scene. Try not to get beat up, but also say "That's just unacceptable. I don't agree with you just because I'm White."
>
> Anyway, it's been a real eye-opener living in the South. People don't think that – when we go back to DC and tell people the things that people say, and they're like "It's not like that anymore." I'm like, "This happened last week."

Like Meg, Ava's states that familial influences and then increased cultural exposure were most influential in shaping her views of human diversity:

> Ava: There are actually two things. Growing up, it was my parents. Their views were very narrow-minded, in my opinion. Once I got into probably about middle school or high school and I started making friends with people that were not like me, that really started shaping my attitude toward [diversity] more just because I got to know people that have different backgrounds that were Spanish, Black, White. It didn't matter. I guess the more people that I met since I've gotten older has really shaped my views.

Researcher: How did that work when you were meeting people that were different from you? Did you have a predisposition about them?

Ava: Yeah, I did. I took a lot of what my parents said and used that as judgment toward people until I got to know them. I was doing football training when I started high school, so I got to know a lot of the guys that played football. The majority of them were black. We hung out a lot, and they ended up being one of the main groups of people that would take up for me because I was really small in school, so I got made fun of a lot. They always had my back. After that, I got really close to a lot of them.

Ava's experiences suggest that increased cultural exposure could have a similarly positive effect of decreasing prejudice and bias for other PSTs.

Matt offers a divergent but related account of how his earlier views of diversity have changed:

Matt: My parents' parents were very upper-middle class. My parents were just very middle-class. Even though we were middle-class, there was this air of refinement and, you know – "be proud of what you are." So, very bland, Anglo-Saxon kind of cultural surroundings.

Researcher: What has been most influential in shaping your views of human diversity, and more generally, people different than you?

Matt: To be honest, I actually got into that kind of stuff through politics. Which happened in about middle school .... In high school, I was introduced to various kinds of texts and things. Anarchism is kind of how I got into radical politics, and from there it naturally led to a lot of reading by socialists and things like that. Yeah, I was kind of introduced to the idea of race and stuff like that through radical politics.

Researcher: So, how were you informed by your readings and your interactions in that area?

Matt: To put it bluntly, I had a very, I guess you'd say, liberal understanding of race. Where it was like, I thought I was very conscious of it, but I didn't really think about it that much. You know, the whole colorblind attitude.

Researcher: Do you feel that learning about diversity from a scientific perspective stands in contrast to learning from the more liberal, colorblind perspective of diversity?

Matt: Oh, yeah. To me, the difference is with the colorblind attitude, it's really ignoring the problem. Whereas, from a scientific standpoint it's understanding issues. So I think it's a totally different approach to it.

Research suggests that Matt's earlier views of race which included the so-called "colorblind attitude," or as presented earlier, what King (1991) refers to as "dysconscious racism," is indicative of a pervasive, yet false sense of understanding of human diversity to which many teachers subscribe. In an attempt to foster a sense of equality by disregarding racial differences, the colorblind individual makes no such distinctions. While seemingly good-intentioned, ignoring race also elides social disparities that often disproportionately affect racial and ethnic minorities, and can therefore produce a similar effect as deliberate discrimination (Choi, 2008; King, 1991).

Sara's familial influences and cultural exposure were somewhat different. She attributes her perceptions of diversity to rich family interactions, her extroverted personality, and her experiences while attending multiple schools:

> Sara: My cultural background is very family-oriented. I come from a big family and we would always get together for some reason or another. It was always a close-knit family. And even though we had tons of issues when we got together, we all still had fun together. We still did things as a family, even though half the time we couldn't stand each other. Sometimes you wanted to say, "I'm never coming back!" and then, the next event, we were all back there again.
>
> Researcher: Do you think that may have contributed to your outlook, both on your own schooling and what you intend to do when you teach?
>
> Sara: Yeah, I think it had a lot to do with it, even though there weren't a lot of new people added to my family until I got older, it was still – it's like my family kind of puts you through a 'trial' when you first come around them. They are their worst when new people come around. It's like they're testing them. They want to see, "Can you really handle this family?"
>
> And as you show them that you're okay – you can handle them being a little bit crazy, it's like you were always there; you're included in everything. And my friends, when they did come around my family, they were like, "I love your family. They're so great." And I'm looking at them like, "Are you crazy?! Have you seen these people?!" But it works for them, and even though I didn't realize I was doing it at first, I was doing the exact same thing.
>
> Researcher: What has been most influential in shaping your views of human diversity?
>
> Sara: I think it's because all the schools I went to, they were always diverse .... It was never the same majority at every school.
>
> Researcher: And was that elementary, middle, or high school?
>
> Sara: It was all three, because I moved around a lot.

> I'm a really open person, so everywhere I moved I always had different types of friends. It was always fun to learn about them. I had one friend, who was from France. She was the reason I wanted to learn French, because at her house, they only spoke French. And she would teach me words so that I would be able to know what they would talk about.

Aside from her personality, Sara's experiences contrast with those of her classmates in terms of hers and her family's acceptance of others. However, like her peers, her experiences speak to the notion that increased levels of cultural exposure can have a positive influence on how we perceive individuals from diverse backgrounds.

The instructors provided greater context on the importance of education to cultural exposure. Tamela makes the following connection:

> I think education is really important, and I think with more education, the more you can recognize and relate to human diversity.

Tamela notes lifelong learners are more favorably disposed toward human diversity due to the desire to understand more about the world around us. In relation to this study, Tamela's comments congeal the importance of realigning education with critical conversations.

These narratives suggest that while teachers' negative perceptions of diversity and students from a different race, culture, or ethnicity (RCE) than their own can be influence by family, they also suggest that cultural exposure as well as the determination to see the world differently can help redress the cultural dissonance created by these influences. However, analyses of the data indicated that sociocultural influences and exposure to science also played a role in understanding REC, as well as scientific thinking.

Sociocultural Influences and Cultural Exposure

The narratives contributing to this emerging theme implicate one's sociocultural background and experiences (i.e., schools, communities, and religious beliefs) as factors that could potentially influence their willingness to use science to address human diversity. Some participants indicated that sociocultural beliefs and/or lack of exposure to science and diversity might also influence teachers' understanding and receptiveness toward racial and ethnic differences as well as scientific thinking.

Extending the notion of cultural exposure in the previous theme, Kara explains how her sense of cultural awareness was impacted by her schooling and community:

> In terms of diversity, I guess in [my hometown] there's not as much diversity. At my high school, they really put a lot of importance on diversity in school, and I think that's why I really loved that high school. When you walk into the common area, they have a flag from every country that every previous student has been to. I think they have a flag from every country in the world.
>
> Even the difference moving from the Christian school and moving to the public school. The Christian school teachers don't really – they don't really teach you anything about diversity, but there's not many diverse students in the school. But that was something that I really enjoyed with learning about – all the different cultures and everything – when I went to public school.

Allie and Bertie describe somewhat similar sociocultural influences:

> Allie: Probably, where I grew up. Yeah and most of my family they're just kind of the same, but where I live it's not very culturally diverse. It's pretty much the same thing with everyone.
>
> It's a predominantly White area. There's not a lot – and it's very Christian. Very, very Christian. Obviously, there was a few other religions, but it was predominantly Methodist and Catholic. That definitely, I think, affected my views on everything. I didn't really ever venture much outside of [my hometown]. And then most of my family is from Florida, and it's kind of the same where they're from so it just kind of carried over into [this state].
>
> Bertie: Not very diverse. I mean, pretty much I've been around the same people my whole life, I guess ... people the same race and ethnicity and background as myself.
>
> I mean, I've never had hard feelings toward any other culture or background. Even growing up in a small town and going to school [where] there's like five people of a different race than myself, I've never seen anybody treating those people badly. They never seemed like they felt that they didn't fit in. But that's obviously from an outside point of view. I don't really know how they would feel.

Alexi adds:

> When I was growing up, I was very sheltered. Pretty much, I did not experience so much with diversity. Meaning, I went to school, but when I came home – there was a life in school. Then, when I was at home, I was strictly only attending everything Russian. I did not have many friends. What affected me the most was as soon as I started going to college. That's when I started experiencing more and meeting a lot of different diverse people with different backgrounds. That's when I started getting into relationships. I was told

> a lot of times that I was really sheltered, and I just did not know. I can't blame my parents for that. They did the best that they could. That's when I started becoming my own individual and experiencing all I could with other people. I think that's my biggest impact, is when I started going to college.

Having been an educator for over 20 years, and as a current teacher educator, Instructor Tamela draws from her own experiences about biases in the classroom:

> Teachers come in with some type of a bias, and sometimes they don't even know that. A lot of people don't even know they have a bias walking into a classroom. And I think we need to educate our educators with that. And I really – I do think that's really important. One of the things that I stopped doing a long time ago was – and I tell my students this – you know, when somebody comes in the classroom and says, "I need somebody to go to the book room and get books." Who do you call on? Call on all the big boys! So, what are you saying to the females in the class, or the 'puny' boys in your class? You know? So, I stopped doing that. I said, "Well, you know, I'll let these people go." Now, I might have to send seven instead of four. But I tried to stop what was the bias between male and female and so forth, like that. And for grading papers, a lot of times, all the teachers chose all the girls to grade papers. So, again, I stopped doing that. It's like, "Boys can grade papers; girls can grade papers." But I think that if we don't recognize some of that in ourselves … and that is just an underlying bias in that example that we can correct in the classroom.

Not only does Tamela confirm the implication that, however benign, teachers bring personal biases into the classroom, she also makes the important point that instructors would do well to be more self-reflective to address such behaviors. Adding to her remark on reflective teachers, she states:

> I think the place to start is really with – if you're talking about education, anyhow – is starting with the teachers themselves, going out into the classroom, and the unintentional vibes they give off and the unintentional behaviors that they reinforce or don't enforce. So, from where your research is coming, if you can start to identify some of that, then we move that into the education field of educating teachers – I think that'll help start with that, anyhow.

Instructor Tamela's comments suggest that developing a critical awareness of self is an integral aspect of effective teaching.

To varying degrees, these examples of sociocultural influence provide context for understanding the challenges that some PSTs may face when trying to reconcile learned behaviors when teaching in diverse settings. In addition to the above, religious beliefs emerged as another factor contributing to this dissonance, particularly with one's exposure and receptiveness of science.

Religious Beliefs

As mentioned in Chapters 1 and 2, religious ideologies have long been at odds with science. It is a deeply conflicting relationship that has been maintained for centuries. The narratives emerging to form this theme suggest that religious beliefs may work against understanding human diversity from a scientific perspective. In her account of the sociocultural impacts of schools and communities, Silva suggests that religious ideology resulted in limited exposure to science, as concepts such as evolution run counter to theological teachings:

> Unfortunately, going to Catholic school for part of my life, science didn't really have much of an impact, because it wasn't, it wasn't focused on. It was really lightly glossed over, and also it incorporated a lot of religious thought. So I know now that the science I got in elementary school when it would have impacted me the most and draw my interest is very, very inaccurate, unfortunately.
>
> It was a bigger part [after moving to the United States]. It was focused on a little more, but because I wasn't exposed to it when I was younger, which is when I would like really grasped an interest for it, I didn't really care for it. So, unfortunately, the fact that science wasn't as present in my elementary schooling meant that I just didn't care for it as I went on to school in the States either.

Silva's comments are grounded in research on international assessments, such as PISA and TIMSS, which indicate that, by Grade 8, students' positive attitudes appear to decline (Mullis et al., 2012; Osborne et al., 2003). Silva's remarks suggest that sociocultural influences and religious beliefs may contribute to this decline.

Similarly, Allie notes the influence of religion on upbringing and education, and how it conflicted with her understanding and application of scientific knowledge:

> If you have a closer family, I feel that really would affect it because the culture of family the more they are going to push you and the more that they're going to want to see you achieve and be successful. I think that has a lot to do with it and not only that, but your practice in your own beliefs. I was raised in a Christian household.

> I would get really uncomfortable in biology to talk about evolution and things like that. It was just like a concept that I couldn't really understand. I just have never really taken to science. And I'm not sure that's why, but it was something that I didn't feel learning it. I didn't want to learn it because it went against my beliefs and it was like that.
>
> That was when I was younger, but now I'm more open to the ideas of different things. Now, I can separate my beliefs versus other people's beliefs and be like, "Okay, well this is a possibility, and this is a possibility, and there's actual evidence behind this," and I can wrap my mind around actual things like theories. When I was in middle school I couldn't do it.

Jess' remarks echo the shift from somewhat constricting religious beliefs toward more scientific thinking and its potential application in schools and society:

> I grew up in the Catholic Church, and as I got to my teenage years I started to rebel a little bit and learning about evolution and all that. I became a vegetarian. The more I learned that science and knowing that humans share half their DNA with bananas, you can't really tell the different kids – boy, girl, black or White, Mexican, whatever they are – you can't really judge them on what they look like. But through their personality and their character.

Similar to Popper's (1963) use of critical rationalism, as well as Kuhn's notion of scientific revolution, it would appear as though Silva, Allie, and Jess experienced a "paradigm shift" of their own resulting from the introduction to and consensus regarding new scientific ideas. Kuhn (1962) argues, for example, that discovery and scientific consensus that decenter normative thinking engenders new scientific ideas. Popper (1963) suggests that reaching this point requires critical assessments of otherwise "rational" information. In this sense, though in different ways, Silva, Allie, and Jess' exposure to and greater understandings of scientific knowledge about human diversity appear to have outweighed previous convictions used to describe such phenomena that could not be falsified.

However, as Ava suggests, espousing a scientific perspective does not necessitate abandoning one's religious beliefs:

> I think science is the base of – I'm a very religious person. I read my Bible every morning. I have a strong relationship with the Lord. But I also believe that science is pretty much the foundation that a lot of this country is built on. I think anything that we do can be looked at from a scientific point of view.

In contrast to Silva's, Allie's, and Jess' acceptance of science, Sara is wary of biases embedded in information presented as scientific knowledge and suggests that science itself might require further examination:

> Sara: Regardless of what people say, I think science is biased. Because everybody's view on science is different, and it's their view .... So, you might understand their view, but you don't understand the whole view.
>
> Researcher: What do you mean "the whole view?"
>
> Sara: One person's opinion on how humans tend to be ....Whether you use science or not, it's always biased because of that person telling you ... it's not a world view of it. I can't get a majority opinion from just one person.

Sara's remarks suggest that she is not entirely convinced that a scientific point of view is always trustworthy. Though, considering the legacy of institutional discrimination and racism in science discussed in Chapter 2, including pseudoscientific practices such as eugenics, phrenology, etc., her skepticism is warranted.

Summary

Taken together, these narratives suggest that the dissonance between one's sociocultural disposition and level of exposure to science could be a delimiting factor that impedes the use of scientific perspective to make sense of the world, as well as explaining human diversity. They also suggest that increased exposure to science can have a positive impact on how one understands the world, irrespective of their sociocultural beliefs.

The narratives associated with this theme indicate a need for greater understanding of human diversity in TEPs, and point to important issues of which educators should be aware when using such an approach. These challenges underscore the potential contribution of CPSP in TEPs, as well as K-12 settings.

From the responses above, it is clear that various social and cultural factors play a large role in shaping how prospective teachers' views prior to entering the classroom. Given that many of the nation's teachers are White and middle-class who teach in increasingly diverse classrooms along with unexamined views of race and ethnicity, cultural disconnections are likely to result in teachers in rather damaging misconceptions of the students they serve (Delpit, 2006; Delpit & Dowdy, 2002; Howard, 2006; King, 1991; Swartz, 2003). However, the question remains as utility of science for teaching about diversity in TEPs, to which Silva offers a brief rejoinder alluding to its possibilities:

> Given what I know now, it's been a better, influential part of understanding that where people come from and the fact that we're really not as different as we seem like we are. I believe that the only really difference between people is region, because the region determines the language you speak, the culture that you have, the customs that you have, everything that you do, but biologically and scientifically, there's no difference.

This comment speaks to understandings about human diversity afforded by scientific inquiry that can help confront stereotypes and prejudice in schools and society. The next theme addresses larger societal understandings of race, ethnicity, and culture (REC).

"Big Picture" Understandings of Race, Ethnicity, and Culture

From analyses of participant narratives emerged the notion that using science as a means to understand the world facilitates "big picture" understandings of REC. This theme not only serves as an interconnecting feature among the other themes, but also resonates with previous chapters which suggested that systematic investigations and applications of science can offer greater insight about natural phenomena and, indirectly, social phenomena (AAAS, 2010).

Seeing the Big Picture

In the context of the study, narratives suggesting that science facilitates understanding is particularly relevant when teaching about human diversity as CPSP is itself intertwined between the natural (e.g., genetics) and the social (e.g., race, ethnicity, and culture). Donovan et al.'s (2019) research supports these important conversations in science disciplines, especially concerning genetics discussions. Much as the following narratives allude, it is in this way that science can provide what might be considered a "big picture" perspective of a world that all humans share. This theme includes participant responses regarding the insights afforded by science about the natural as well as the social world. Furthermore, whether or not they maintain a scientific mindset about such phenomena, participants often attached a relatively high level of importance to science for what it can illuminate about the world around us. Sara explains how science shows a bigger picture of the world, but also contends that scientific consensus would be necessary for it to be relevant:

> Sara: I think that would be more of a big picture type thing. Because it's almost like a puzzle when you put a bunch of people together and they're all contributing information. Yeah, they're biased, because

> it's coming from their opinion, but when you put it together with a bunch of other people, you see a bigger picture. Then, you get a better understanding.
>
> Researcher: Do you think that type of scientific information would be valuable for preservice teachers?
>
> Sara: Yes, I think that would be valuable. Because when you're teaching you have to be able to see the big picture, and you have to be able to show your students that bigger picture so that they're not just focused on one person's idea – they have several peoples' ideas and they can put together their own conclusion.

As mentioned, in nearly all cases, participants reported some level of deepened understanding about the world as a result of scientific endeavors, many of whom also emphasized the relative absence of opinion when using science. A prominent view in this regard was that science provides factual evidence to support its claims and the implication that such "proof" could facilitate understandings about the world. For instance, Allie states that:

> People are more apt to listen to what you're saying if you have some scientific proof behind it. If you just come out of the gate and you're just like, "This is what I think, but I have nothing to back it," then it's not going to have an impact. But I think having scientific data or anything that can show people, and be like, "This is it. This is real." I think that can definitely change a lot of what people think.

Allie's remarks suggest that scientific evidence could help allay doubts and have a positive effect on how individuals think about and respond to issues of race and ethnicity in the classroom. This is a point that becomes more evident in related themes; however, in what immediately follows, participants describe their views of how science can be applied to understand fundamental aspects of the natural world.

Understanding the Natural and Social Worlds

Some participant responses illustrated the informative power of scientific investigations for understanding natural phenomena, others extend these understandings into the social world. These views surfaced mainly in relation to activities that address the notion of "authentic science" which involves applications of scientific inquiry and problem-solving in real-world scenarios to understand different facets of everyday life (Crawford, 2014; NRC, 2012). For example, for Matt, science is informative about everyday occurrences often taken for granted:

> I mean, it's all just awesome to learn about ... like we were talking about in the classroom, you know, how you can bring science in, truly

a lot of things, to help us understand it better is really important. You know, how often we don't think about how – I think you were saying in class – about how we don't think about the physics of a car rolling down the street. But, it's nice to be aware of that kind of stuff, and to take in mind that you're driving this two-ton machine down the street.

Meg shares the influence of authentic science in her life and comments on applying scientific information toward understanding the natural world:

> We're a very curious family. We would always go to museums all the time. Growing up we did – my dad … I guess I was always just really curious now that I think about it. We went to a couple museums and I was a kid, but when I became a parent the museum became really important, and we bought lots of science books, and we did lots of science experiments. Just the natural world – my kids play with – we made more baking soda volcanoes than I could count …. Currently, the whole climate debate to me, there's too much time wasted on saying "Look at all these climate deniers." Obviously, the climate is changing. Let's talk about solutions. We're wasting so much political energy on climate deniers. It's like finger-pointing. It's just spending time on people who still think the world is flat. It's just stupid to talk about what we know.

Ava offers another account of engaging with authentic science to understand the natural world:

> I remember, probably a year ago, we found a snake in the backyard and we killed it. It's really gross that I'm saying this but to me, as a mom, this was a perfect opportunity to show my kid how a snake digests food. I cut the snake open, and it had two whole birds in its body. So I'm showing my son this, and his eyes are just like wide open. He's so excited to be seeing all this. I know it's probably gross to do that to my child, but to me that's science. You're showing them nature and reality.

For Ava, nature serves as an access point for understanding the world. By involving her son, she seeks to provide him with the same access point from which to develop his own understandings of the natural world.

In addition to "seeing the big picture" of the natural world using scientific data and evidence, many participants described connections between science and the social world. When asked about the influence of science in her life, Ava expounds on earlier remarks and relates science to understanding the social world:

> The biggest part for me is just the simple scientific method. We consistently learn it throughout middle school, high school, even

> in college. I just got done taking a biology course, and everything you do – you have to approach it that way. You have to make what they call your hypothesis, in your head, and then you observe. You make records or facts .... To me, most of the things are observation. Just watching the people. That's why I don't feel like being able to look at somebody's file before they come into a class – I don't think that that's a smart choice. When I was a freshman, I wasn't focused on school. I was more of a talker. But then, the older I got, I got more excited about college, and I started focusing on my grades. So, if somebody would have looked at my file – maybe they did – my freshman year, they may have prejudged that. I think it's better to watch people and treat them according to their actions and not what's hearsay or what's been told. My senior year, I had the same social studies teacher I had my freshman year, and she gave me terrible grades. She was the only class that I ever got bad grades in. It was almost like she saw me, and she just already had this idea in her head of who was coming in her class and did not even realize that I had grown over three years into a different person.

Ava's narrative suggests that authentic scientific observation would not allow for such myopic assessments. Instead, a teacher would acquire more data by studying students' behavior to avoid prejudicial assumptions.

For Silva, who spoke very little English when she moved to WSV from a small, Puerto Rican town, technological advances of science that gave rise to social media outlets were most influential in her understanding of the social world:

> I think where it [science] helped me most is social media, because I would see people discussing things and get interested in them and then find more about it and then learn more about what it is that I was experiencing without actually being a part of it. I guess because I couldn't be a part of it and I didn't know how. So I learned a lot about different issues, like racial issues in the United States, and a lot of problems – a lot about diversity and religion. Definitely the diversity and the different cultures that people have ... I didn't know there were so many different people.

Silva's remarks about using social media to engage with and better understand the social world are redolent of the impacts of social media on various uprisings around the globe. As mentioned in Chapter 2, nearly 90% of Egyptian and Tunisian youth who were surveyed in March 2011 indicated that they used social media to organize protests or to raise awareness about social and political issues (Huang, 2011). Domestically, recent research demonstrates over two-thirds of U.S. citizens report the use of social media for political or social organization (Pew Research Center, 2018).

Sara offers a different view of technology and the role of science in understanding the social world. When asked what science could tell us about race, ethnicity, and culture, she states:

> Sara: I think on some levels, it kind of eliminates it, because when it comes to science and technology, the only time you notice race is when you change the language on any devices you have.
>
> When you use a computer, you can get help on how to set it up in different languages. But, other than that, we start out the same way with technology, regardless of what language you use. So, the cultures of different races don't come into it, when it's about science and technology.
>
> Researcher: Do you think that any of those applications of science and scientific knowledge could contribute to better understanding race relations, multicultural communities and ideas?
>
> Sara: I think technology can bring that into your home, because learning about different cultures and how they do things, you don't have to go out and find somebody else to show you. You have it right there in your house. All you have to do is look it up. You don't really interact with it too much, but you can still experience it through technology.

Sara makes interesting observations regarding the interplay of science, technology, and culture. In her view, technology serves as a neutralizer of culture as well as an access point. Together with Silva's comments, these narratives suggest that science – by way of technological innovation – is a function to understand REC and, perhaps, mitigate negative effects from the lack thereof.

To an extent, Instructor Tamela agrees that technological advancements add tremendous value to education:

> I really value the advancements in technology, and how our world is becoming – you know, the health, and resources – there's just so much that technology's done. I love teaching – I only had a couple years with the SmartBoard – I was teaching social studies class and we went to the Amazon River, and you could zoom right down on it and actually see the river and the creatures and stuff from the satellite. That was just a wonderful learning experience, a visual learning experience.

However, as a longtime educator, she also suggests that the overuse of technology can negatively affect one's ability to understand and interact with the social world:

> And where I think we're 'missing the boat' is – I think all of that is wonderful, I think that needs to be a part of our classrooms and moving forward with that – but somehow, I think we've got back off all this social media and all of the 'social' that our kids and students do through technology. Somehow I think we have to bring that back to actual socializing with human beings, not socializing through technology. And to be honest with you, I'm not really sure how to do that.
>
> And I know I talk to parents all the time, and they say, "Well, they won't do this ... they won't go out and play, because all they're doing is on the phone or on Facebook ..." and these are elementary and middle school kids, a lot of them.

Instructor Tamela's comments illuminate a different aspect of the conversation regarding the use of technology for understanding. On one hand, technology can open up the world and facilitate "big picture" understandings of, in this case, natural phenomena that one might otherwise get to experience in such detail. On the other hand, without self-moderation, technology can be restrictive and isolating, thereby closing one off from the natural and the (human) social world.

Understanding Race, Ethnicity, and Culture

In several cases, participants' comments regarding enriched understandings of the natural and social worlds using scientific knowledge converged into this ancillary theme, which leads to more specific connections of the natural and the social worlds. In this sense, natural and social phenomena are combined using science to provide understandings of RC, such as natural (genetic) human similarity versus social (racial) categories and stratification. For instance, when asked what science might contribute to more inclusive schools and society, Matt spoke of "broadening" the way people view and discuss race through the scientific study of humans:

> I guess when dealing with race and ethnicities, you have really fantastic things that are coming out of anthropology about how humans developed, how cultures developed. So, I think especially that field can help immensely to understand the backgrounds of cultures and how they developed, which can help you immensely in understanding areas like that. It's just broadening people's vision of the world, their understanding of the world. And overall, that's going to help you learn in other areas, too. You know, that kind of background information. I mean, it helps immensely I would say.

Matt suggests that by investigating humanity with a scientific lens, one can reach deeper understandings of REC in ways that expanding their view of the world, perhaps, in ways that promote tolerance and inclusion of difference.

For Meg, using science to understand REC and "where people come from" has important implications for education as well as society:

> I think it has huge value. There are lots of things that you just need to know is an educated human being .... I think there's a lot of value in understanding where people come from if your job is going to be educating people. If you know the origin of humans and how everybody looks different and why we look different, you're so much more informed when you have to take a class on diversity, which is "Don't say these words." You know what I mean? A lot of those diversity training classes are just "These are all the things you can't say." Instead, this is all the science. These are preconceived notions. I thought that [race literacy] quiz you gave about the genesis of laws was really interesting. I think there would be huge value in having that course be a required component.

In highlighting what she views as the substantive contributions of a CPSP approach, Meg's narrative indirectly relates to Matt's earlier statements about colorblindness. Her comments reflect literature that suggests that colorblindness is often imbued within a TEP's diversity courses which tend to focus on more superficial elements of human diversity, omitting deeper issues of racial inequality (Brown, 2007; Cochran-Smith, 2003; Hickling-Hudson, 2004; King, 1991).

In the interview, Bertie stated rather unequivocally that she did not care for science and does not think about it very often. However, she was no less able to make connections to its potential role in understanding REC:

> I believe that science is very important. Just because I don't think about it doesn't mean it isn't important. But at first, when I saw this study, I was trying to figure out how the two [science and race] were related. But, definitely after listening to your presentation in class, I really can see the correlation between them. I mean, I had no idea that there were no different genes between the races. I think when you tell people that it definitely hits home and makes them realize that we're really not so different. I think when you tell students that it makes them more accepting of people that aren't like them, and more ready to talk and interact and feel more comfortable ... in today's world, everybody has come up with their own conclusions as to why everybody's so different, and they're really not accurate at all when you look at it from a scientific point of view.

Along with "seeing the big picture" (i.e., understanding the natural and social worlds), these narratives illustrate how science might be used to facilitate deeper understandings of REC and also address its potential to help address challenges PSTs might face when teaching in diverse settings by understanding human similarities.

Human Similarities and Race as a Social Construct

This concept emerged in similar discussions regarding the informative potential of science for understanding REC, and particularly race, both scientifically and socially. Some of the responses regarding the use of science to understand REC highlighted human similarities as an integral component of doing so. Ava points out, for example:

> I think, many times, we're making judgments on color of skin and preconceived notions. It's almost like science could prove that we are all the same, and it doesn't matter who you are and where you came from. We all need to educate ourselves and be better to the world into each other. However science could go at that – that's how it needs to go.

Jess adds that acknowledging human similarities does not exclude one's particular family customs or cultural traditions. Rather than dissolve cultures, her comments suggest that such understandings help create space for them to exist:

> I think it helps. I think if we teach people from a young age that that's true – that we're pretty much all the same; that our differences are based upon where were born and who are families are – I think we can also have our different traditions and cultures. But we can share them with every one instead of being segregated – we are still really segregated. We can come together and learn about each other and not go, "That's wrong! You're doing it wrong!" and just accept what people do.
>
> I think it is important just because in the generation that my grandmother grew up in – she's White and she talks about hanging out with a little black girl. And she talks about it like they're not all just kids hanging out – there are these White kids and then there are these black kids, and then they just happen to mesh together. And it was like, "Well, yeah, you're all just kids." They weren't taught that they are the same, they're all just kids. And if we teach our generations and the next generations that there aren't really any differences, I think they will grow up to not have racism – there just won't be any point.

Silva agrees:

> I think it would make it easier because even as we set ourselves apart from each other, it made me feel very alone. So, having that information to rely on you could be like, "Hey, they might be from somewhere else, but we're in the same situation, and we have the

same origin. We're going through the same experience. We're really not that different. It's okay."

Alexi concurs and connects her understanding of human diversity and similarities to the information gleaned from her science classes:

> Alexi: Taking those classes, I feel like it shows that we're all the same.
>
> Researcher: Do you think those concepts can contribute to better race relations in schools and society? If so, how?
>
> Alexi: Absolutely, they do contribute a lot. A lot of students view other students based on appearance or physical appearance. That's because they haven't really developed their true identity, and they're just kind of lost out there. They judge other people by their appearance. If they were to actually know how we're made up on the inside and what defines them as people – not only how we're made, but how we are different with our attitudes and everything else – as you intertwine that together and show that we are the same, even with our emotions, but we are different. I don't know how to say it. I think that it's important for children to know that science aspect with our DNA. That should be emphasized in their learning.
>
> Researcher: Do you think that would help teachers reach their students more effectively?
>
> Alexi: I do. But I feel like the way I see it, I think we need to focus more on how much we are similar. For example, we have the same amount of bones in our bodies. We're all the same, but depending on how much melanin we have in our skin, we see that minor stuff that we see.

Alexi's relationship with science leaves little question in her mind about what science can reveal about human diversity and the similarities therein. In addition, given that she would ultimately be a science teacher, Alexi was insistent about developing her ability to implement science and human diversity in her future classroom. She states, "This is a big thing if I'm teaching chemistry! How do I implement that?"

Instructor Jonathan is similarly convinced of human similarities:

> Human diversity – it's the science. As a science student, I have always believed that all humans are the same, are from the same origins. So, for me, it's never been a question. It's a 'no-brainer,' because I believe in science.

A trained biologist, Instructor Jonathan maintains a clear perspective of the science that justifies human similarities. Extending the discussion, Matt notes that race is not biological but rather a construct of the mind:

> I'd say the most obvious point would be that science disproves the idea of race – that it's a social construct that we constructed – and really any kind of political and economic ideology or whatever. You know, when you look at the 1800s and you have things like eugenics and all that kind of stuff. And today, science has come out that shows that that was ridiculous – things like that. And specifically with race and culture, again, it helps disprove the social construct, and with things like gender – things like that. So, it's very important. At the same time, I think it has its limitations, too. Where, specifically, if you were to teach that race is a social construct, I think you'd also have to combine that with the idea that, while race is socially constructed, the oppression based on race is very real.
>
> So, it needs to be combined with that, in my mind, to really be effective in talking about race and things like that.

Matt's response is indeed grounded in scientific fact, and also illustrates that deeper understandings of race help put into perspective the realities of social stratification based on racial categories. His comments suggest that science reveals race as this product of disillusioned social opinion. Similarly, Allie references a point made during the class discussion to express her views about what science reveals about race: "I think, as far as race goes, you said it very well – didn't we just make that up?"

Silva shares her view of what science contributes to understanding this concept:

> I think it [science] can tell us everything about race.
>
> Science can boil down a human to their DNA. It can tell you everything, because that's what we are – we're DNA, essentially. And science has the ability to tell us what our DNA is composed of now, and DNA can be trace back to anything.

Silva's remarks speak to a fundamental "truth" about understanding the natural and social worlds in relation to REC, which Instructor Jonathan helps explain:

> When people are given wider scopes of knowledge, which science affords in its most elementary way, everything begins to make sense. And that is the reason why science is very compelling because science forces you to your knees and makes you obey the laws of nature.
>
> When you understand wider scopes of disciplines, then you are actually approaching the original intent of what we call science, which is nature study. Anybody who observes nature understands sociology. I mean, what is sociology?

> It is the study of communities, actually. In ecology we do communities, don't we? Ecology is communities. "Ecos" is literally the study of the home –
>
> I think that's really, in a way, that's why biology becomes interesting, because biology forced me to wonder more about everything. So then, all of a sudden you're interested in sociology … anthropology. It's all there. It's really the science of nature. And, in a way, that's basically what you are saying: to what extent does the science of nature engender an understanding and more embracing of everybody.

Instructor Jonathan posits an interconnectedness of the natural world that through scientific inquiry helps understand aspects of the social world. For example, that that race is a social rather than natural (biological) phenomenon. He then explains how in the absence of such scientific thinking unexamined truth claims help construct stratified social categories (e.g., race) and subsequent inequities in a society:

> When you look into all societies around the world where they are more metaphysical than science, they are not progressive. I mean, look at India. And it's very interesting, and a rather ironic twist. Where you have some of the best so-called gurus on earth, living side-to-side with some of the poorest people on earth. They have created something akin to human filth, which is really sad. And they have created these hierarchies of human existence, that kind of thing, which doesn't make sense. Those kinds of things don't make for human progress. So, in a way, people who decide by mental propositions which may be faulty, at best, and impose those for realities, all of a sudden creates societies of inequities. So, you have the "untouchables" and all those – you go to China, you have the emperors; you go to Europe, you had the gods and the laws. And you have all these systems. And then, all of a sudden, the science comes in to explain advancements – you know, where rain comes from and the 'rain-god' dies. Because, all of a sudden, we know the water cycle. Okay? We discovered the emperor has no clothes. So, we don't need a king or queen, so they lose their positions. And so, what is interesting is really, how all these rather silly human creations, which are intangible and unscientific then crumble under their own weight because they are artificial human creations and do not stand up to scientific scrutiny.

The foregoing remarks ultimately suggest that science facilitates understandings about the natural and the social worlds by broadening and deepening perspectives regarding human diversity (i.e., REC) using scientific knowledge and principles. With these new understandings of REC that diminish social categories such as race, PSTs are better equipped to create space for equitable and inclusive learning environments.

Summary

Narratives in this theme describe how science can be used see to a "bigger picture" of the natural and social worlds that facilitates deeper understandings REC. Donovan et al.'s (2019) research similarly suggests that genetics discussions with both adults and students help to foster greater sense of diversity awareness. Participants suggest that science not only expands one's views of the world-at-large but also provides insights for making distinctions between natural and social phenomena such as genetics (biological) and race (social). For example, one of the supporting concepts for this theme, *human similarities,* includes natural phenomena such as genetics which can be examined scientifically. Whereas, for the other supporting theme, *race as a social construct,* one could use genetics to understand that, scientifically, stratifying society based on race is the equivalent of social stratification based on eye color. One implication is that by making such distinctions, prospective teachers would be less likely to conflate these discrete concepts and therefore resist debilitating assumptions about students based on racial categories.

Racial Anxiety and Apprehension versus Meaningful Dialogue

Participant narratives express apprehensiveness about discussing or confronting race-related issues, while others suggest that using science to understand and discuss human diversity can help create reduce racial anxiety and encouraging more meaningful dialogue about such issues.

Meg shares her apprehension about addressing race-related issues or incorporating strategies similar to those in this study in her future classroom:

> Researcher: Do you feel that with the information that I presented we could have continued that discussion? Did it give you the impression that we could start to unpack this a little bit?
>
> Meg: It did actually. It did feel like that. I'm a fairly educated, experienced person – listening to your lecture, and it was really informative and really interesting. But there was part of me that was thinking, "I don't know if I could be this frank if it was me instead of you giving the class."
>
> Because I'm a White person, or maybe it's because I'm older – I'm another generation older than you – I don't know. I'm not concerned having this conversation with you on the phone, but some kid might take something wrong that I said ... I would be a little nervous. What if I said something the wrong way, or it started a conversation in my classroom that somebody said something and then I lost my job?

Although Meg was receptive to CPSP, her responses suggest that teachers may be reluctant to use such an approach to avoid discomfort or for fear of losing their job.

Ava shares Meg's concern but feels that using verifiable information to negotiate such discussions makes the challenge of apprehension a surmountable one:

> To be honest, the way that girl in class was talking about – I'm just more nervous about saying or presenting something in the wrong way, where it's not been that way, but in my opinion it shouldn't be taken in a bad way. To me, that's the most concerning thing. I hang out with people from Puerto Rico, Panama. I have black friends, White friends, and all of that. I get along with all of them really well, and we're all really close. I'm just more concerned with somebody misinterpreting something I say.
>
> I've always wanted to teach. A lot of the fear of teachers right now is losing their jobs because of a parent suing them or a student saying something. You're eliminating one of those things. You're taking it and saying "Here's a good way to talk about the situation" without putting yourself in it or putting anybody else in it. You're looking at the facts.

Upon graduation, Bertie will join approximately 80.1% of the teachers in this country who are White (NCES, 2017), and another 76% who are female. However, she was apprehensive about the prospect of being in the minority at a culturally diverse school:

> I've always pictured myself teaching in a school very similar to the one I went to. Now that I'm here, I'm thinking, "What if I student-teach at a school downtown that is full of people completely different from me?" And I don't know, I guess you just don't address the differences and you have a classroom discussion, call on people equal amounts – I don't know. I feel like if you talk about it like you're comfortable about it, then they will feel comfortable about talking about it, too.
>
> I feel confident in the fact that I will be able to treat everybody in the classroom equally, but it does make me nervous to go into a school where I might be the minority and to have students looking at me, and judging me, and me being on the reverse side.

Though, when asked about social justice and addressing human rights issues in the history classroom, she suggests that despite this apprehension, confronting and overcoming any such discomfort may be a necessary aspect of teaching history:

> I mean, both of those are very important aspects of history and you can't escape the past. You have to address it. And I think it's

> important to address either way. Just because it makes you uncomfortable doesn't mean it's not important and it doesn't need to be talked about.

Silva's apprehension stems from uncertainty of whether she could adequately implement CPSP strategies in her future classroom:

> I would feel comfortable once I learn how to do it properly because I wouldn't want to give misinformation. I wouldn't want to say something that's not true, but definitely I would feel comfortable using it, especially because if I were to go into teaching English as a second language.

She also expresses reservations about teachers' receptiveness diversity regarding their ability versus their willingness to incorporate CPSP into their instruction:

> Circumstantially, I think it'll contribute to understanding to the people that care. But from my point of view, there are just people that don't care to understand race, ethnicity, or culture and the presence of science isn't going to make a difference to people that just aren't bothered to learn.

Silva raises an important point about the tenability of CPSP and its potential to more broadly affect society. In relation to the preceding narratives, her observations highlight a fundamental concern with any type of reform: the extent to which the public will gravitate toward new ways of thinking. One implication is that, despite a teacher's ability to confront issues of diversity in the classroom, their discomfort and apprehensiveness may ultimately dictate the willingness to do so.

In contrast, extending her earlier remarks about using science to understand REC, Ava connects scientific understanding to desensitizing discussions of race:

> I think if we could get the things like race and stuff in that same way to use the science to make it not such a touchy subject but more of an interesting and fun subject, to educate would be awesome.
>
> If there wasn't so much time spent on the nervousness and the tiptoeing and all that, there's so much more time that could be spent letting these kids learn more things that had substance, that could better their life. So, when they get out in the world, they're not like, "Be careful, there's a Spanish [sic] person. I can't say anything about that. Let's not talk about that." That's ridiculous to me.

When asked how and the extent to which the information in the class discussion using CPSP was a viable for understanding and improving race relations, she adds:

> The data, being able to use that and the way you presented it kind of made me feel more comfortable; that if I ever have to deal with a race issue coming up or cultural issue that you don't have to have an opinion about it. You can use facts. Then it's not your way, it's the way that's been proven. I think that's the best way – just finding that kind of information that you presented, and just using that to the advantage of, like you said, proving where people come from, what their relation is. People can either turn a blind eye, or they can educate themselves. That's the best way to going to happen – if we get people to stop looking the other way and start paying attention to the facts.

Ava's narrative suggests that understanding human similarities and using scientific facts may help reduce teachers' anxiety and the earlier mentioned apprehension about issues related to race and ethnicity, thus encouraging and facilitating more meaningful dialogue about the same.

When asked if she could imagine a scenario in which CPSP could be used in an English classroom to facilitate discussions that involved sensitive topics such as REC, Allie responds:

> I know when you're teaching 18th and 19th century literature, or when you're learning it, there's a lot of different ways that people portray how they feel about those times. And we just had a class … we were learning about Young Goodman Brown, and they were talking about his life … and it turned into a huge argument. It was not a fun class. I think if people had a better understanding of how other people will think and come to those conclusions, or if they thought about that and how people are going to think differently, I think the class may have gone more smoothly had those issues been addressed.

In suggesting that science would facilitate such dialogue in other classrooms she makes an interesting distinction between mathematics and the humanities:

> I think it would definitely be more helpful in discussion classes. I think I'd be able to implement it more in an English setting, because it's so open thinking and analytical, versus a math class where everything is like, "This is the answer, and this is what you're going to get. No matter how many different ways you think you're going to get it, this is how you're going to get it." And definitely in art, or music, or literature and things like that, you can express freely as far as like what you're thinking and all that.

> If you help your students be more comfortable with the culturally diverse classroom, I think that opens up a lot of discussion. I think it makes everyone more comfortable and it's easier to learn in that environment you feel comfortable to say what you think. I think helping students understand that – and it would be boring if everyone looked the same – helping them understand would be a lot better.

Allie suggests that using science to reduce race-related anxieties in class discussions creates a more inclusive learning environment. And yet, while she acknowledges the potential of science to create an inclusive atmosphere in the classroom for English and the Arts, because they allow for discussion and free expression, she imagines that mathematics is perhaps too rigid to include a CPSP approach.

Bertie agrees that incorporating science in discussions of diversity is helpful and also fosters a sense of equality in the history classroom:

> I feel like I will definitely bring it up when anything about diversity comes up in history. I feel like it's important to remind everybody that just because it happened doesn't mean it's right.

Similarly, Matt believes science could support meaningful dialogue in history classes, as well as much a needed public discourse about race:

> I mean, for me, overall, this is just something important to think about teaching history. So, it's good – it's nice to talk about it, I guess.
>
> Yeah, without a doubt … there is no racial or ethnic dialogue really in this country, I find. So, if it's not being talked about, people aren't going to pick up on it or be aware of it, unfortunately. It should definitely be something that people go through.
>
> Allie agrees with Matt, and suggests that education is a good place to start:
>
> It would definitely help to introduce in society, but I think it really starts with education and in the education system because everybody goes through that and obviously we need it in our society. I think the best way to reach the mass would definitely be through the education system just because you have to at least go until 12th grade.

Meg extends the discussion with remarks on the educational and social utility of applying scientific concepts to understand human diversity:

> It seems like people are really afraid of talking about race. Hopefully, we're moving to a place worth...less difficult to talk about it. I think probably it's not just White people. There are probably a lot of other cultural groups that are free to talk about race. There has to be a

> way of constructing the lessons where people can feel really safe talking about them.
>
> I think helping people put the focus on our cultural differences and things like the lack of opportunity that makes us different, not our biological differences. So just putting the focus on where the difference is. Making assumptions just based on skin color – that's just for ignorant people ... I feel like the more we can talk about the science, then we can focus on lack of opportunity and real issues that make us – not just different, like kinds of economics, but just different – everybody wants to be with people that have similar interests. It's just when you make value judgments on it that people get hurt and killed, like that kid in that gas station, because he was listening to – I forgot what that guy called it – jungle music or whatever. That's what we have to fight against.

Meg's comments are germane to the larger implications of CPSP in schools and society. One of which is that using science to facilitate race-related discussion would create space to address other social inequities as well.

Successful Teachers in Diverse Settings

These concepts correspond with participant narratives that describe what it means to be a successful teacher in diverse settings and either directly or indirectly address general qualities of responsive teaching. In line with previous themes, there was a shared sense among respondents that, irrespective of the subject area, using science to teach about diversity was a feasible, if not essential, aspect of developing truly responsive teachers and fostering inclusive learning environments. Given the implication that CPSP could facilitate meaningful dialogue about such issues, these descriptions illustrate the types of responsive educators CPSP can support. For instance, Matt suggests that responsive teaching in the history classroom might include empowering students from all backgrounds to critically engage history using their own voice and cultural lens:

> To be a successful teacher in a history setting, for me, it would be addressing the particular racial or ethnic group in history to make sure that they get a sense of connection to it. As compared to, especially US history – it just focuses on White figures and things like that. Also, in history, a sense of voice. Because, to me, history is more of an argument. So, I would want to encourage students to interpret history through their own tools and things like that – to come up with their own conclusions about events. I would just want to try to foster a sense – where they could bring in their racial or ethnic background and use that to interpret history.

Kara describes how her high school teachers were responsive to students from various backgrounds and the impression that made on her:

> I think what's been most influential is that my teachers at [high school] put a lot of importance on having respect for everyone in the classroom, and making sure everyone was treated the same, no matter what their background was. And I think I've learned from that a lot.

Alexi is very passionate about learning and teaching science and feels that education was a good fit for her because she can use that to inspire students. When asked what draws her to education the most, she responds:

> What draws me the most? I do want to help people, and I love working with children. In becoming a teacher, I feel like I can put passion in somebody's heart. I'm very empathetic, and I really want to work with kids. I feel like teaching would be the correct way to go.

She is also confident about teaching in diverse settings and points to language as an entry point for cultivating an inviting classroom and connecting with students.

> Researcher: Describe your thoughts about teaching students that might be culturally different than you – racially, ethnically, etc., and how confident do you feel in teaching those students?
>
> Alexi: I know whether it be in any district I teach in, there are going to be racially and ethnically different students. I feel like I would be able to do that. I don't have any doubts that I won't be able to. I never really – I mean, I've thought about that, but I have not really considered it. I don't feel like I will be incapable of it. As long as I don't exclude anyone and just make everybody feel equal.
>
> Researcher: You said you wanted to "make everybody feel equal." How might you go about doing that in your classroom?
>
> Alexi: Definitely through language. Even something as simple as saying "we." Just having everyone feel as one. Knowing everyone's ability when I get to my students since everyone is different. That shows through their background, their home life. Just integrating a positive lesson plan and how all the students can interact is one, I feel like adds a great way to have everyone feel equal.

Guided by empathy for others and her passion for education, Alexi points to the use of language and differentiated instruction in ways that promote a sense of equality as responsive modes of teaching. Using CPSP to facilitate meaningful dialogue about sensitive issues supports creating such a learning environment.

Sara comments on being responsive to culture in the math classroom. She conceptualizes responsive teaching as a form of culturally differentiated instruction:

> A lot of people say math is the same no matter who you learn it from, who teaches it, where they come – math is the same, but it's not. Because the technique and the application of how math is taught is different from culture to culture. And when you have a diverse classroom, sometimes you have to go with whatever culture that student knows, because that's how they learned the basics. You're able to teach them your way, but unless your way can connect to the way they learn, it's not going to help them with the subject.

Similarly, Allie contends that responsive teaching includes being adaptable, open minded, and avoiding prejudicial assumptions:

> You have to adapt to everybody's needs, and their strengths and weaknesses, whether that be cultural or academic.
>
> You definitely have to keep an open mind. Going into it close minded, children are going to think the same way. And even adults – you can't think all adults are going to think the same way. You have to get to know before you assume. You can't put everyone in one category. If you do that, then you are limiting yourself to a very minimal amount of knowledge.

Jess adds:

> Each interest, you relate to something in the world. In science, you relate to biological evolution. And in history, America itself has come from different countries – and all of our different histories – everything has something to do with people. So, you would have to know not to judge people before you taught your students anything.

In agreement with Jess and Allie, Bertie states:

> You need to know about different cultures and different traditions. Get to know more about where these people come from, and their home life and what they're used to and what they respond best to, in order to be able to communicate with them and teach them as best you can.

Instructor Tamela draws from her years of experience as a teacher educator and behavioral specialist to provide context to the discussion:

> I think to be really, really successful at teaching, and at learning, but especially for students to learn, especially students that are

> challenged to learn, they definitely need the structure but they also need some free-flow, some creativity to it. They're not just going to sit there and memorize in rote, they're just not going to do that, whether they can't do it or they're not going to do it because they're not motivated to do it. So you have bring that kind of artistic creativity into the classroom along with their structure and their content and try to be – you know, every day is not going to be a dog and pony show, it's not – but like I tell my students, my [UT] students, at least if you can try to incorporate something with them – a creative lesson, as much as you can and come in with little tidbits, or just keep surprising them each day with something.

Silva takes successful teaching another step further when describing the type of responsive educator she intends to be. Interestingly, she was the only participant to explicitly suggest that responsive teaching also occurred outside of the classroom:

> As an educator, I would probably talk to students outside of class and make sure they're okay if they're understanding, and see if there are any resources that I can accommodate them with, but during class definitely, I wouldn't treat them any different than I would treat the other students because they are capable of doing everything that American students are capable of doing. It's not fair for them to feel like they're not, only because they are culturally diverse.

Jess also spoke at length about the need for teachers who can inspire students to stay in school, and her subsequent desire to be such an educator:

> I don't think that education in this country is what it should be. I think it's sad that kids don't get to experience – like my school didn't experience – any diversity. And there are kids that have to go through security to get into school. And there should be teachers that they get see that inspire them to do better in their lives. And then yesterday we had the speakers from Uganda who spoke about children walking six kilometers or eight kilometers to school and being so eager to learn through terrible circumstances. And I think anybody who's in the education program especially should learn about what other people have to go through in order to understand how to teach the people who have to struggle in life.
>
> My dad didn't graduate high school because he didn't have anybody to inspire him. And he could have done a lot more with his life have had he finished high school and gone to college. And there are a lot of students that don't see the point in high school or college. And maybe I can't make them realize that, but maybe I can help them stick with it until they graduate.

Jess' concern for education was evident throughout the interview. Her comments suggest that education supports social mobility and that responsive teachers are those who help students understand its inherent value.

Along with narratives in the *Racial Anxiety and Apprehension versus Meaningful Dialogue* theme, the foregoing descriptions of successful teachers in diverse settings illustrate the types of responsive educators CPSP can help promote.

Summary

This theme discussed the need for meaningful dialogue about issues of REC in schools and society, and that the ability of science to provide supporting evidence allows for seeing a bigger picture of the world such that human diversity was viewed more the ironic parallel of human similarity that reduced race-related anxieties than the contentious and divisive network of uninformed diatribes it has become. One implication is that CPSP strategies could inform such debates within and beyond (science) classrooms and schools, and into the public discourse.

Furthermore, the supporting concept of successful teaching in diverse settings highlights aspects of responsive teaching that align with subsequent narratives which discuss the applicability, accessibility, and adaptability of CPSP in TEPs and K-12 settings. Together with the views expressed throughout these findings, this theme suggests that CPSP may well be a substantive approach to various forms of responsive teaching in diverse settings and a starting point for meaningful discussions about human diversity. It also suggests that having sound applications and pedagogical practices are essential tools for achieving this goal.

Applications, Strategies, and Impacts

This major theme emerged from narratives related to the potential impacts of applying a CPSP strategy in schools and society to understand and discuss issues of diversity. All respondents suggested that using science to learn about human diversity in education would be beneficial. Most suggested that it would assist in teacher preparation, while others suggested ways this approach could be used in TEPs or in K-12 settings. Some participants offered strategies for using science in ways that directly align with, supplement or extend CPSP, in general, or in their future classrooms. In addition, throughout these narratives are implications that using science to understand human diversity is relevant to teachers of all academic areas. Some suggested that CPSP may have a cross-discipline accessibility that would allow it to be seamlessly integrated into any subject area or that it would complement differentiated instructional strategies. Bertie states, for example:

> On the very first day of class – whether you're going over history that day or not – when you present the syllabus and everything, I feel like definitely that's something that needs to be addressed. Especially in history classrooms, because you have to have a lot of discussions.

Matt speaks to the implications of using a CPSP approach in teacher education:

> It's like the thing we worked on in class, that "race literacy test." I don't think a lot of people know about these kind of things, so it's important to get that kind of information out there in the day-to-day .... And in the classroom, I would think it would just help understanding particular students. Without any sense of cultural awareness, any sort of problematic thing they [teachers] might see as behavioral issues .... So, having science behind you can help cultural development. You'd be a lot better at managing your students by understanding your students and where they are coming from, which I think would help you develop a much better relationship with a student.

The "Race Literacy Quiz" that Matt (and Meg) refers to was developed by California Newsreel and the *Association of American Colleges and Universities* was administered to students at the beginning of the class discussion in this study (California Newsreel, 2003). This "quiz" asks a series of questions related to science and history as well as racial, ethnic, and social awareness. Its purpose was to highlight the importance of knowing about issues that disproportionately affect certain racial groups. The narratives from Bertie and Matt suggest that it quiz might be a good way to implement CPSP.

Teacher Education and K-12 Settings

As earlier suggested, the value of science beyond the science classroom was a common view among participants in discussions of the utility of science for learning about human diversity. When asked if they could envision a CPSP approach being implemented in teacher education, the consensus was that it would indeed be beneficial:

> Kara: I think it would be a good idea for all teachers to take a class on science with diversity itself, because the science aspect of it proves it true – there's no arguing it really. I think it makes other people believe it – it makes it more believable.
>
> Alexi: If you're specifically in teacher education classes, I most definitely do see that as being important. That's what we need to learn in order to apply that when we would become teachers. It's the reality we're going to have to face. We need to know now and be educated with that and further address that in the future. It's just little things – not little things – the big thing we don't think about really. It's really

> important to know that now and to keep that in mind. Like I said, I haven't really thought about that. It is very important to have us learn about that now in teacher education courses.
>
> Silva: I would it would be beneficial for any education classes. I think it would be beneficial for any education classes that – in the specific instance that I'm thinking of, it would be really beneficial in the intro to special education class, for example, because a lot of people like to set themselves apart from kids or students who need special education, for a lack of better words. But, regardless of circumstance, we're the same. I think information like that would be very nice to incorporate and to have just the kind of get rid of the stigma and to stop people from setting themselves as far apart as they do from people that require special education.
>
> Meg: Yeah, I think it is really valuable. I think that would be a great course for teachers. I think if you're developing curriculum it has to be 'real.'

These narratives indicate the applicability of CPSP in teacher education. Participants consistently reference the value and benefit of science as tool that informs through its ability to deconstruct well-worn ideas about race and human diversity. The implication is that CPSP would make a valuable contribution to TEPs and therefore support the development of effective teachers who were more critically aware of the imposition and harmful effects of social norms such as race.

Participants were also asked whether using science to address diversity in TEPs would be more fitting as a standalone course or as an additive to existing courses. Most said a standalone course might be most effective; however, all participants thought that, at a minimum, it would be helpful as an additive to multiple courses:

> Allie: I think probably a standalone class – there's so much to learn on it. You can't just plan to do what we got out of it and then expect people to actually take something from it and use it. It would definitely have to be an entire class where you sit down and you help future teachers, and let them know that this is what you got to do, because it's more embedded .... An entire course of the science of it alone I think it would help tremendously.
>
> I think it would be helpful either way. But if you just plan on doing it for a week or two, it's going to stick with some people and with some people it's just going to be brushed right off. If they do it a week before they student teach, and if they go to the classroom and they utilize it for that week, it may stick with some and, for some, it may not. But if you have an entire course and if you're practicing and applying it, it's definitely going to leave an impression on people .... It would definitely still be helpful just to have that, but I think it would be more effective to have the entire course.

Ava: I definitely think it's something that needs to be taught, whether it's as a full class on its own, or just in a teaching training program. Whichever it is, it's obviously a touchy subject for people right now, especially everything that's going on in the world and the way people are being treated. It's something that needs to be addressed before we get into the classroom and we're like "Uh-oh, I am not prepared for this."

Matt: Oh, yeah. Most definitely. I imagine it would be so much better if it was just a specific course, not just for teachers, but everywhere.

Bertie feels that adding science on a class-by-class basis would be sufficient:

> I think we've worked it into this first class pretty well and I think it could just be mentioned in each education class you have to take. I don't know if this should be a whole class in itself.

While a minority suggested that it would be best as an additive, all participants agreed that this approach would make a valuable contribution to teacher education.

When asked about other educational applications, several participants indicated that CPSP would be useful in various capacities in K-12 settings as well. Allie states:

> I think anywhere that anybody can learn about it would be – anytime that you can take that opportunity. Because teachers, now, they're not – this isn't something that they were even considering 10 or 15 years ago. I think getting people and sitting them down, no matter what age, I think it would still be helpful, definitely, to help older teachers into a newer generation and a new way.
>
> So, [with] the idea of races, it would definitely help to apply science and the history of the people that you are trying to talk about to reach an understanding as to, in my case, having to explain to them what they just read.

Allie suggests that new and experienced teachers would benefit from CPSP. She also mentions how she could use this approach to help explain readings in her future English class that include sensitive topics involving REC.

Silva views CPSP as a way to address differentiated instruction:

> I think it should have a role in the school level because when it comes to understanding culture and society we often look towards Language Arts and Social Studies and those two. Although they are a very good component for understanding different cultures and race in society on their own, they don't appeal to every mind that the students would have, like some students leaning more towards

> the sciences, some lean more towards the arts, and some lean more towards languages. Does that make sense? I feel like students that lean more towards sciences don't have that kind of understanding to fall back on because science is not focused on much to understand race in society, but I think it would be very important component if it gets incorporated properly.

Silva's comments suggests that a CPSP approach would be beneficial to discuss REC which, she notes, science hardly does, but also to appeal to different types of learners. Hers was one several ideas for CPSP strategies that were generated by student participants in this study.

Strategies for Implementation

Kara enjoys learning about other cultures and views such knowledge as a valuable part of the educational experience. She imagines variety of ways to infuse a CPSP approach in a K-12 setting:

> Yeah, one hundred percent. I think it would be cool for a high school course just to have a whole course on diversity and learning about different cultures. Even in English classes, with the 9th grade literature or British composition, there should always be some sort of learning about different cultures and diversity even within that. Obviously, in literature, you are learning about different cultures, but shining light on every culture ... you can always do that within whatever course. Even if it's a math course learning about different ... calculating the differences in genes or just whatever. No matter what course or specific topic, you should always include something ... so that the students can see the differences and the similarities with other students, technically.
>
> It's just common knowledge, really, being able to share that. I think even if I was an English teacher, it would be an interesting prompt for students to write on.

Although Kara is an aspiring English teacher, she mentions several subject areas in which CPSP could make valuable contributions in K-12 settings. Notably, though not to dismiss or belittle Allie, one of the other aspiring English teachers in this study, Kara found a connection to mathematics (i.e., "calculating the differences in genes"). In addition, she offers very insightful ideas including a high school course on diversity, as well as using CPSP topics for writing prompts in her future English classes.

Ava, also an aspiring English teacher, explains how she might incorporate CPSP in her future classroom which includes strategies used in this study. Specifically, a video entitled "3 Ways to Speak English" that shown to the class at the end of the CPSP discussion to humanize and to contextualize the scientific discourse presented throughout. In it, poet and educator Jamila Lyiscott confronts misperceptions of intelligence in society based on how a person looks or speaks, using spoken word poetry (Lyiscott, 2014). Ava's comments also address the cross-curricular accessibility of CPSP, as well as the importance of differentiated instruction and responsive teaching:

> Honestly, I would probably use your technique. Give some facts, and maybe show some videos. Everybody doesn't learn the same way. You've got to have the ones that you can talk to and explain things to and show that, and then you've got to have the ones that you move by videos – like the girl that was speaking. I got chills when she was talking. With things like that, you can get to different people. Some people are creative. Some people are just there to be there. You've got to hit everybody in their sweet spot.
>
> With English, it's a lot of breaking things down, pulling them apart. You're dissecting words and sentences and trying to see how I can relate to you or how you can use it later on. If it's a book that is cultural, ethnic or racial, that's probably how I would handle it. I would just pull it apart, give the facts about it, and let them see what it is. As questions come up or discussions start, then I would probably approach that as it came.

Ava acknowledges that reaching the most students will require various strategies, to find their "sweet spot" and keep them engaged. She also appears to equate analytical aspects of English with the deconstructive elements of CPSP.

Cross-Discipline Accessibility and Differentiation

Similar to Ava's and others' previous comments, some participants brought up the cross-disciplinary accessibility of CPSP and differentiated instruction, and at times, the interdisciplinary features of education. When asked if were possible for to easily integrate CPSP into their lessons without being a science teacher, participants on the whole suggested that was a possible, and many felt that it may be necessary. For example, Bertie suggests that cross-curricular teaching and differentiation helps galvanize important topics in the learning process.

> I definitely thought it was neat, like, when I was in high school and I would be in history class and then all of a sudden we were talking about a book that I was reading in English at the same time or talking about something I had learned in science the year before. I always thought it was neat when they all kind of ran together … I think it's important to tie all these different subjects together. It makes it more important.

Silva believes that any teacher can relate other subject areas to the one they taught, they need only the will to do so:

> I think with proper training anybody can do anything. I mean, any teacher that would be willing to do it could do it. Teachers can find a ways to incorporate … like, chemistry teachers can find ways to incorporate ice cream into lessons. So, if you can incorporate dessert … I remember very distinctly in middle school we were trying to find a way – no, in high school I was taking a Chinese class and we wanted to order Chinese food, so we decided to have a lesson on chopstick etiquette. But it was just an excuse to order Chinese [food]. So, teachers can find ways to incorporate anything into their lessons if they try hard enough, and with proper training and proper information – with the proper tools in their 'toolbox' – any teacher can incorporate this information about race and ethnicity and origin into any of their lesson plans.

For Jess, the cross/interdisciplinary nature of education is a requisite to a well-rounded college experience:

> I think another thing is that people get stuck going, "Well, I'm an English major. I don't want anything to do with science."
>
> That's the whole point of college, is you don't just do English in college. I'm not just in Spanish. I'm in astronomy and I was in math. And that's for a reason because we need to have a – what do they call it? When you're well rounded. You need to be well rounded in your life. So, if everybody is just well rounded because they got a good education, then we'll all might just get along.

Some responses included more generally the idea that CPSP strategies would be valuable for all teachers.

> Allie: Yes, definitely. It'd definitely be useful as far as – I mean, I can't strictly rely on my knowledge of literature to get through class.
>
> Meg: I think it should be for teachers. It should be for anybody teaching.
>
> That quiz that I took I thought was really interesting … I would take that to the first meeting of my French class or the second week of my French class.

Bertie's reprise suggests that CPSP would be a good way to start any class:

> I think no matter what subject you're teaching, it's important to bring up on the first day of class that everybody's going to be treated with respect and everybody's equal. And you could definitely bring up that scientific point of view to show that everybody is the same.

Although content mastery is essential to high quality teaching in any setting and vital to preparing a highly-skilled workforce, as Allie suggests, simply being proficient in their content area is not sufficient for teaching in diverse settings (Gay, 2002; McConnell, Parker, & Eberhardt, 2013; Wang, 2013). Bertie and Meg similarly note that using CPSP to discuss human diversity at the beginning of the school year would be valuable whether it was used in a history or a foreign language classroom.

Summary

This theme presented narratives related to various applications and strategies for using science to discuss human diversity in education. Overall, the responses support the notion that CPSP strategies are valuable for TEPs and K-12 settings in part because of the cross-discipline accessibility of CPSP and more generally the interdisciplinary nature of education. They also suggest that rather than impinging on "non-science" teachers, incorporating CPSP strategies supports the differentiated learning for both students and teachers. And further, that its application in diverse classrooms use may also promote responsive teaching. Though, similar to her earlier comments about apprehension, Silva's remarks suggest that applying CPSP strategies is contingent on teachers' willingness to respond to the needs of their students.

Chapter Summary

This chapter presented the findings from the analysis of interviews conducted with ten (10) PSTs and two (2) instructors to understand the potential utility and potential impact of using science to teach about diversity (CPSP) in TEPs. Participant narratives suggest that much can be learned about human diversity through science, including insights and strategies for cultivating more inclusive learning environments, both in TEPs and in their future careers as educational professionals. Despite apparent challenges that also emerged from the analysis, the overall tenor of the themes similarly echoes the utility of a scientific lens for understanding human diversity. Participants point to various applications of CPSP that deploy scientific inquiry and scientific knowledge to facilitate understandings of REC. In this way, the narratives presented in this chapter support the premise of this book to determine the utility of science for teaching about human diversity in TEPs, and to provide PSTs with related insights that can inform their future practice in positive and inclusive ways. Chapter 6 discusses findings presented above using a critical postmodern framework of science, diversity, and education.

6 Critical Role of Science in Teaching Diversity Discussion, Recommendations, and Implications

Previous chapters outlined the significance of the study, situated this research among relevant literature, described the research design, and presented the findings. This final chapter discusses the findings using a critical postmodern framework of science, diversity, and education. The research questions for this book are: (1) *What is the role of science in teaching about diversity to undergraduate preservice teachers at an urban research university in the Southeastern United States?*; and (2) *How might this approach impact preservice teachers across disciplines in terms of raising greater critical awareness and influencing inclusive classroom practices?*

The chapter begins with a discussion of the theoretical framework, which is then used to discuss the major themes and subthemes of the study. The major themes were: *Critical Challenges and Sociocultural Dissonance*; *"Big Picture" Understandings of Race, Ethnicity, and Culture*; *Racial Anxiety and Apprehension versus Meaningful Dialogue*; and *Application, Strategies, and Impacts.* Findings of the study are also connected to critical postmodern science (pedagogy), which serves as both the theoretical lens and method of investigative inquiry and instruction.

Theoretical Framework Revisited

This section revisits the study's theoretical foundation, which merges critical social theory and postmodern social theory to unveil oppressions and unexamined truth claims in schools and society. The result of this convergence is a critical postmodern framework of science, diversity, and education applied as a pedagogical strategy to teach about human difference. This is known as *critical postmodern science pedagogy* (CPSP). With respect to its focus on human emancipation and decentering absolutes, this composite framework stands as a critique not only of social and institutional oppressions but also of any abuses levied in the name of science. Furthermore, when applied as a pedagogical strategy, CPSP becomes an incisive tool for understanding race, ethnicity, and culture (REC) in ways that support responsive and egalitarian teaching.

In support of PST preparation, CPSP likewise engages past and present social issues and determines their implications. Namely, how institutional apparatuses such as law enforcement provide a space wherein historical racial discrimination manifests as the disproportionate killing of at least 9 Black women and men under the pretext of self-defense, public safety or law enforcement over the past 6 years. The deaths of George Floyd, Breonna Taylor, and Ahmaud Arbery, Eric Garner, Tamir Rice, Sandra Bland, Trayvon Martin, Jonathan Ferrell, Botham Jean, Korryn Gaines, Atatiana Jefferson, and countless others who died too soon, speaks to the need for more diversity and inclusion dialogue as well as policy changes and systems of accountability. These killings and ensuing racial discord and civil unrest suggest that, while human rights in society is indeed further afield than in the 1950s and 1960s, there remains a need for greater understanding about what it means to share a community humanity, how that is possible, and the implications of critically aware cohorts of teachers using such a perspective in the classroom. Therefore, given the complex and inextricable connections between schools and society, unearthing and discussing REC issues using critical theory and postmodernism are important to carry out the deeper analyses necessary to develop and promote inclusive teaching practices.

The critical theory underpinnings of this framework serve as an investigative tool for questioning sociocultural patterns and assumptions tethered to societal prevalence rather than critiques of their effects. One effect of the prevalence of these often unexamined patterns is the tendency toward exclusion of those whose appearance and/or values do not align with the dominant or hegemonic class, which remains anchored in Eurocentric cultural values and ideologies at the expense of non-Whites. Various organizations, such as the National Association for Multicultural Education (2019) and National Science Teachers Association [NSTA] (2020), have since ostracized science disciplines for omitting diverse narratives. The theoretical perspective of postmodernism rejects and disbands grand narratives or absolutes of any kind (Lyotard, 1984). In this way, a postmodern perspective offers a disconnected view of sociocultural phenomena and insights relative to their social and historical loci; though, not in search of truth, as such.

Rather, when used toward egalitarian ends, postmodern social theory allows for an "empathic detachment" from which those things presumed to be true can be surveyed and then deconstructed. Though not entirely divorced from the self-assigned task of critical theory to more directly upend social institutions that perpetuate the subjugate and objectify the individual, a postmodern perspective is concerned with decentering the toxic network of intersecting oppressions from within the very structures that confine freedom itself by reconstituting the intents and functions of social discourse and action (Collins, 2000; hooks, 1990; Horkheimer & Adorno, 1972).

Critical theory's macrosocial view of society levies disarming attacks on overtly oppressive discourses and tactics and disinters those that remain hidden. The goal of this metaphysical battle is to empower and transform the individual from *Object* to *Agent* of change such that they might escape the shackles of false consciousness and restrictive banalities of hegemonic discourse that governs social action, thereby freeing oneself from the alienation therein (Gramsci, 1971; Horkheimer, 1982; Marcuse, 2001). A postmodern approach, however, views such attempts to critique *and* escape material conditions using normative analyses as cognitive futility. For, as Lorde (1984) puts it, "the master's tools will never dismantle the master's house"; that is, "What does it mean when the tools of a racist patriarchy are used to examine the fruits of that same patriarchy? It means that only the most narrow perimeters of change are possible and allowable" (pp. 110–112). A postmodern perspective seeks instead to unravel, de/construct, and re/arrange these discursive formations, thus providing more nuanced understandings of the complex interplay between truth and power (Derrida, 1978; Foucault, 1980). And yet, without strong moorings in criticality and grounded understandings of the social world, which critical theory provides, in its opposition to grand narratives postmodern theory risks dissolution into fantasy or perverse relativities of morality (Baudrillard, 1994; Lyotard, 1984; Steinberg & Kincheloe, 2010).

Such is the intersection and "synergistic" relationship between critical and postmodern theory as it is used in this book (Steinberg & Kincheloe, 2010). The strength, then, of a critical postmodern framework of science, diversity, and education rests in the collective ability of these two theoretical perspectives to detect, confront, disassemble, and then re/construct oppressive discourses and conditions. Thereby, critical postmodern frameworks provide enriched understandings of the world, which can be used to confront illusive truth claims and false consciousness on material as well as conceptual levels in schools and society such that one might reclaim their "ontological vocation" of human agency and liberation (Freire, 1970; Horkheimer, 1982; Marcuse, 2001). A critical postmodern science theoretical framework thus becomes an important lens for analyzing the findings of the study.

Critical Postmodern Science Analysis of Findings

This section discusses the findings presented in Chapter 5 in relation to the critical postmodern framework and CPSP instructional method.

Critical Challenges and Sociocultural Dissonance

As mentioned in previous chapters, due to an increased number of science, technology, engineering, and mathematics (STEM) professions over the last several decades and the current lack of qualified individuals to meet this demand, teacher education programs (TEPs) have become an integral nexus for preparing a diverse and highly skilled workforce to support an eminent shift to a technically advanced knowledge-based economy in the 21st century (Carter, 2008; Cochran-Smith & Zeichner, 2009; Darling-Hammond, 2010; Gordon, 2009; NRC, 2007a, NRC, 2007b; Riegle-Crumb & King, 2010; Wang, 2013). Also mentioned was that while student populations across the country grow increasingly diverse, approximately 80.1% of their teachers are White, and 76% are female (NCES, 2017; Swartz, 2003). However, as CPSP helps demonstrate, this largely "White, monolingual, middle-class, female" cohort, alone, is not the issue (Banks & Banks, 2004; Brown, 2007; Milner, 2006, p. 344).

As participant narratives in the *Critical Challenges and Sociocultural Dissonance* theme indicated, it is a troubling combination of social inequalities that pervade, especially, underserved schools and communities comprised racial and ethnic minorities, and the perpetuation of which through unexamined hegemonic cultural values transmitted through social institutional channels (Delpit, 2006; Delpit & Dowdy, 2002; Gay, 2002, 2010; Howard, 2006). This was evident in student responses related to several of the supporting concepts for this theme, including influential factors such as familial influences and cultural exposure, sociocultural influences and exposure to science, as well as religious beliefs. Meg's and Ava's respective accounts of the contemporaneity of racial discrimination in the South help illustrate this point. Meg, a native of Orange County, California stated:

> Truth be told, I never heard anybody say the "N-word" out loud until I moved to [this state]. People just dropped it in the conversation. I literally cannot even say that word out loud. I can't even believe – It took me off guard.
>
> Anyway, it's been a real eye-opener living in the South. People don't think that – when we go back to DC and tell people the things that people say, and they're like "It's not like that anymore." I'm like, "This happened last week."

Ava was born and raised in the South. Growing up, she shared her parents "very narrow-minded" views of diversity and "took a lot of what [her] parents said and used that as judgment toward people until [she] got to know them." It is clear that Ava learned prejudice at home, and from the people she trusts most. Interestingly, Meg discussed similar learned behaviors that provoked her fear of African Americans that she maintained into adulthood:

> My grandparents had owned that apartment building in Compton, and then they started building public housing in Compton. My grandparents sold their building because of what was happening. They must

> have been talking about what was happening to that neighborhood and stuff and blaming it on black people. I was picking up on that...
>
> I grew up thinking that I didn't have any racial bias against African-American people until I moved to D.C., and I realized that I had a fear when I saw African-American people. I recognized it. It took me a while to recognize it, but I didn't realize I had it until I spent time with African-American people. Then I had to be like, "Okay, you're being stupid. This is ridiculous."

As narratives in Chapter 5 suggest, although some of the participants' views have since changed due to increased cultural exposure and while the CPSP approach in this book appeared to allay some discomfort associated with discussing issues of race, Ava remains "nervous about saying or presenting something in the wrong way" in her future English classes while Meg is apprehensive because "some kid might take something wrong" that she said. One implication is that preservice teachers (PSTs) from all across the United States may bring similarly culturally embedded experiences into the classroom, but perhaps without the necessary critical awareness or methods to reconcile static sociocultural ideologies with the desire for personal growth.

From a critical postmodern perspective, these and related comments reflect an increased sociocultural dissonance beckoned by a rapidly changing demographic landscape and anachronistic views of race imparted by family, or what Popper (2002) describes as a conflict between "conviction" and "inter-subjective testing." However, it would seem that through cultural exposure and increased awareness, both Meg and Ava were able to confront truth claims about racial categories presented to them as absolute, and continue to replace such ideas with enlightened understandings of human diversity. Thus, in line with the literature on teacher education and the CPSP framework, the intersection of increased diversity in schools and society, cultural mismatches between students and teachers, along with unexamined truth claims borne out of dominant cultural values, rightfully remain one of the foremost concerns in teacher education (Howard, 2006; Milner, 2006). Further suggested is that in the absence of critical awareness of human diversity or tools, such as CPSP, with which to confront misguided ideological convictions, many PSTs will indeed enter diverse classrooms unprepared to address some of the issues they are certain to face in the years to come (Cho & DeCastro-Ambrosetti, 2005; Howard, 2006; Ladson-Billings, 2000; Milner, 2006; Swartz, 2003).

"Big Picture" Understandings of Race, Ethnicity, and Culture

In addition to contextualizing the critical challenges above, participants also noted that applications of science to understand both the natural and the social world facilitate *"Big Picture" Understandings of Race, Ethnicity, and Culture.* As mentioned in the previous chapter, the broad notion of "understanding" emerged from many of the narratives, as well as questionnaires, discussions, and notes used in this study as a common thread among the other concepts. This theme includes participant responses regarding insights afforded by science and scientific inquiry about natural and social phenomena that inform deeper understandings of REC. Participant discussions and applications of scientific concepts that facilitated such understandings were a promising aspect of the study.

As noted earlier, starting with the premise of a unified but shifting network of patterns that can be understood through systematic study, (natural) science involves attempts to "figure out how the world works by making careful observation and trying to make sense of those observations" (AAAS, 2010, p. 5). This suggests that similar applications of scientific information can assist, though not fully explain, social phenomena. The larger implication, then, is that applications of science raise important questions about the social world. For example, the oldest human jawbone fossil (~2.8 million years old) was recently discovered in Ethiopia by a research team from Arizona State University. And while older fossils have been found that also link the first humans to an African origin, such as *Australopithecus afarensis*, better known as "Lucy" (~3.2 million years old) and *Ardipithecus ramidus*, better known as "Ardi" (~4.4 million years old), the significance of this find is that it is from the Homo genus, the next closest link to modern humans, Homo sapiens (sapiens), and thus, fills a 400,000 year gap that helps explain human evolution and migration (CNN, 2015). This scientific discovery raises social (as well as cultural) questions that may challenge oppositional beliefs regarding a common human origin in Africa as well as biological evolution. However, a CPSP lens reveals how these findings and context promote broader views of the natural and social worlds, and thus, deeper understandings of human diversity.

In the current study, several participants indicated that science provided a lens for seeing a bigger picture of the world that included individuals from all racial and ethnic backgrounds. Some attributed this expanded perspective to scientific evidence and data presented in the class discussion to demonstrate the extent to which all humans are similar. For instance, that each human shares 99.9% of their genetic makeup with all other humans. The remaining 0.01% representing the phenotypic, or external expressions of genetic coding, such as the melanin content in one's skin which is observed as skin color. Students were also presented with further explanations of how variations in skin color and other external features were the result of adaptations to the natural environment over time as well as proximity to the equator, where the sun's ultraviolet radiation is most intense, which helps explain the variation in color (i.e., darker skin protects against harmful UVB rays).

Some participants were aware that there was such a small difference between humans or other aspects of biology, geography, and anthropology (and sociology) presented in class, others were hearing this information for the first time. However, all participants were able to make broader connections between the natural and social world as a result of CPSP and suggested that such knowledge helped put into perspective not only that humans are similar, but that the conflation of race and biology, then, must be a socially constructed idea, as Matt previously noted, "I'd say the most obvious point would be that science disproves the idea of race – that it's a social construct." Matt's comments reflect the notion that race is a social rather than natural (biological) phenomenon. In addition, such knowledge facilitates understandings and evokes questions regarding the realities of social stratification based on racial categories. One connection of this theme to the CPSP framework is evidenced by the application of science – an almost exclusively positivist, Eurocentric enterprise traditionally used against those who did not resemble an "ideal type" of human promulgated by dominant class ideologies – to deconstruct and recast it as a method to help liberate from racial discrimination both the oppressed and the oppressors (Freire, 1970; Weber, 2009).

In this sense, culture becomes a very important aspect of one's ethnicity. Culture entails all of the social norms, values, beliefs, expressions, and rituals that can be identified with a particular social group. This suggests that while race simply is a social construction that tells us how a person looks, culture serves as the real currency that makes up an individual's identity. Critical postmodern science (pedagogy) thus becomes an important lens for analyzing the findings of the study. Furthermore, due to the sociocultural dissonance that discussing or confronting race-related issues often carries with it, the participants express that using science to understand and discuss human diversity can help reduce racial anxiety and encourage more meaningful dialogue about such issues.

Racial Anxiety and Apprehension versus Meaningful Dialogue

To begin the discussion on the next theme, *Racial Anxiety and Apprehension versus Meaningful Dialogue*, one participant, Bertie, predicts some of her own forecasted nervousness. Bertie notes:

> I feel confident in the fact that I will be able to treat everybody in the classroom equally, but it does make me nervous to go into a school where I might be the minority, and to have students looking at me, and judging me, and me being on the reverse side.

Bertie raises an important point regarding what happens when the teacher becomes the minority in public schools. This can create a lot of anxiety and apprehension in teachers and students. Given that 80.1% of the teachers in this country are White (NCES, 2017), Bertie, who is a White female, will likely experience a racial and cultural mismatch that many teachers struggle with in the classroom. Thus, it is crucial that PSTs have a strong foundation in multicultural and diversity perspectives in education (Mensah & Jackson, 2018; Pride, 2019). In this way, CPSP (strategies) plays an important role in opening up the dialog on a very sensitive yet important topic in TEPs (Ash & Wiggan, 2018). The participants in this study expressed how CPSP helped broaden their views and deepen their understandings of human diversity in ways that support positive and inclusive strategies and classroom practices.

Similarly, the literature on critical pedagogy, teaching and learning, and science pedagogy conveys the importance of authentic and democratic dialog for student achievement (Atwater, 1996b; Barton & McLaren, 2001; Basu & Barton, 2010; Kirmaci, Buxton, & Allexsaht-Snider, 2019). This research demonstrates the possibilities of science in education to assist personal transformation as well as developing the ability to locate and confront social injustice in ways that empower students and teachers to become change agents in schools and society (Bandura, 1989; Freire, 1970; Giroux, 1988, 2005; Kanpol, 1992; Kanpol & McLaren, 1995b; McLaren, 1999). In this way, scientific inquiry and investigations become a tool for disrupting oppressive discourse in schools and society (Atwater, 1996b; Barton & McLaren, 2001; Basu & Barton, 2010; Rahm & Brandt, 2016; Sheth, 2018).

Akin to CPSP, the literature on CRP and science also underscores the need for greater accessibility to science for students from diverse backgrounds (Aikenhead, 1996, 1997; Atwater, 1996b, 2010; Gay, 2002; Hickling-Hudson, 2003; Jegede & Aikenhead, 1999; Nieto & Bode, 2012). This aspect of research on critical and culturally responsive science pedagogy also points to the need for awareness of dominant cultural values and hegemonic discourses embedded in education, including: "the selective inclusion and exclusion of material" in the curriculum, high-stakes accountability measures that disproportionately affect urban schools attended by racial and ethnic minorities, as well as the increasing presence of corporate privatization (Asante, 1991; Barton & McLaren, 2001; Bryan & Tobin, 2019; Hickling-Hudson, 2003; Kumashiro, 2001, p. 4; Lipman, 2000, 2002, 2004; McLaren & Farahmandpur, 2001; Rahm & Brandt, 2016). In helping students restore and maintain their humanity, critical and culturally responsive science pedagogies provide meaningful entry points for scientific inquiry for all students (Irvine, 1992; Ladson-Billings, 1995b; Nola & Irzık, 2005). For CPSP, these approaches represent powerful additives, in terms of leveraging the investigative power of science for teaching about human diversity.

CPSP helps open dialog on sensitive topics surrounding REC in the classroom without essentializing or marginalizing any particular social group. Using science as the basis of analysis helps decenter oppressive discourses in education and also creates a safe space for students to discuss human diversity along a continuum that begins in East Africa. The scientific acknowledgment that all human beings share 99.9% of the same DNA debunks racial, cultural, gender, and religious superiority claims across the family tree. This is an important issue, wherein other forms of teaching about diversity one can go through the process but remain racist, sexist, and/or a religious fundamentalist. In light of these important outcomes, TEPs must respond by providing PSTs appropriate tools and strategies to confront such issues at a deeper level than those currently provided.

Applications, Strategies, and Impacts

Another common view among participants was that a scientific perspective is a valuable contribution in teacher education as well as K-12 settings. Such applications of science beyond the science classroom help PSTs develop deeper understandings of human diversity regardless of the subject area. This awareness can help create an awareness regarding issues that can disproportionally affect minority groups, namely, discrimination and teacher bias. In this sense, teachers' understanding of issues surrounding diverse learners becomes crucial to student outcomes. Therefore, teachers should be cognizant of the strategies they can use to foster diversity and inclusion. As Bertie explains:

> I think no matter what subject you're teaching, it's important to bring up on the first day of class that everybody's going to be treated with respect and everybody's equal. And you could definitely bring up that scientific point of view to show that everybody is the same.

Bertie's comment points to the importance of a scientific perspective of diversity. She mentions that a teacher can use a scientific perspective to show students that beyond race everyone is the same. As such, CPSP and other multicultural and diversity approaches become crucial for effective teaching.

Banks and Banks (2010) argue that multiculturalism and multicultural education are a rejoinder to increased cultural intersections and the subsequent need for "gender, ethnic, race, and cultural diversity of a pluralistic society [to be] reflected in all of the institutionalized structures of educational institutions" (p. 447) through concerted efforts to incorporate the "content, concepts, principles, theories, and paradigms from history, the social and behavioral sciences, and particularly from ethnic students and women's studies" (2004, p. xii). These and other conceptual understandings are integral to addressing educational inequalities that most often affect racial and ethnic minorities. Mediating these disparities can include adapting a curriculum that highlights the achievements and language of non-Whites (Banks, 1993a, 1993b; Banks & Banks, 2010; May, 1999; Sharpe, 2005). As mentioned earlier in Chapter 3, the *Portland Baseline Essays* provide some of the most comprehensive examples of de-colonial curricula framed with historical accuracy.

Increasingly, multicultural education has drawn from critical and culturally responsive approaches to help support and strengthen this mission, and to address the complex interplay of science, society, and diversity (Atwater, 1996a; Cochran-Smith & Zeichner, 2009; Emdin, 2008a; Gay, 2002; King, 1991; Milner, 2006). And because science is still dominated by a male Eurocentric perspective, this literature also addresses the need for a diverse and highly skilled workforce through conceptual and pedagogical methods such as critical pedagogy, CRP, and multicultural science education (MuSE) (Atwater, 1989, 1996a, 2011; Barton & McLaren, 2001; Basu & Barton, 2010; Hodson, 1993, 1999; Hodson & Dennick, 1994). While each of these areas of research has made significant contributions to education and, indeed, to this book, the implications of using science to teach about diversity are virtually absent from the literature. As such, this book is so positioned to contribute to a growing body of knowledge on diversity in education through CPSP.

In a related literature, CRP, which is itself an extension of critical pedagogy, addresses social justice and inclusion by attending to students' ethnoracial, cultural, and linguistic differences in the learning process. Not unlike its critical counterpart, when applied to science CRP demonstrates the importance of connecting subject area content to "the cultural characteristics, experiences, and perspectives of ethnically diverse students as conduits for teaching them more effectively" (Gay, 2002, p. 106). Critics such as Bowers (2005), Gur-Ze'ev (1998), Tubbs (2005) and Rømer (2011) claim that critical and culturally responsive pedagogical approaches are little more than utopian and positivist-leaning variations of neoliberal practices that inherently seek to normalize education such that they conform to universal truths and resist self-examination due to a fixation on material forms of oppression and willful ignorance of abstract consciousness. In this sense, critical science pedagogy may be mis/construed as a harbinger of the very repression it seeks to subvert. However, one implication of CPSP is that it can avoid incapacitating the transformative potential of students and practitioners and create more reflective and responsive perspectives in education (Bazzul & Siry, 2019; Ladson-Billings, 1995a; Laughter & Adams, 2012; Mensah, 2011).

Likewise, with the goal of empowerment, CRP and science view students as agents rather than victims. As such, it seeks to assist students in the personal development of a critical consciousness with which to confront social disparities in their home communities (Barton, 2003; Barton & Berchini, 2013; Ladson-Billings, 1995a, 1998; Lee, 2001). This research suggests that a relevant and responsive approach to teaching science supports high academic achievement regardless of the subject matter or a student's cultural background (Gay, 2002; Ladson-Billings, 2009; Lee, 2001; Lee & Buxton, 2008; Lee & Fradd, 1998). Relating to students' culture, the use of language in schools becomes central to teacher pedagogy and student outcomes.

In examining the role of language in science, Brown (2006) discovered that students from diverse backgrounds found it difficult to reconcile their personal identities in the realm of politicized public and scientific discourse. Addressing this challenge, Emdin (2016) used hip-hop music and popular culture as a political act to teach science to a diverse group of students in an urban setting. In this study, students not only added the contextualized voices that Brown (2006) observed was lacking, but in the space created by this critical approach were able to reclaim their humanity by way of inclusion in contemporary scientific discourse (Emdin, 2007, 2008a). Similarly, Buxton (2010) employed a critical and political approach to science with diverse urban students by having them compare and contrast the quality, distribution, and consumption of natural resources in both poor and affluent areas surrounding their local community to raise student awareness of social inequalities. Donovan et al. (2019) also outlined the benefits of students researching genetic similarities as a way to increase racial understanding. Here, the use of scientific knowledge helped students better understand genetic similarities in the human family tree. These studies suggest that critical science pedagogy is one way to raise awareness of political issues within and beyond the science classroom. And while these critical perspectives and macrosocial inquiries involving science may indeed contribute to the development of anti-essentialist, antiracist education, their impact is often moderated by the challenge of consistent application (Atwater, 2011; May, 1999; McLaren & Torres, 1999).

In this way, CPSP has the potential to re/locate and de/construct unexamined truth claims in schools and society, and particularly those regarding the intelligence and cognitive ability of racial and ethnic minority students (Marks, 1995, 2008; Wiggan, 2007). Drawing from a critical postmodern perspective of science, this book aimed to unveil and decenter oppressive hegemonic discourses by re/framing science as a pedagogical tool and instrument to bring about greater awareness surrounding human diversity and issues of difference. Based on the findings of this study, the following recommendations are being made.

Strategies, Implementation, and Recommendations

The participants suggest that TEPs should have either a standalone course on using science to address diversity, or include it as a component of an existing diversity-related course. While a few students suggested that CPSP should be part of an existing course, the majority of the participants believe that it would be most effective as a specific course in

TEPs. Relating to the issue of having a prescribed course using CPSP in the TEP, some participants also thought that it would be beneficial to have professional development training for in-service teachers on this topic. This would entail having outreach and support or in-service public school teachers to ensure the effective implementation of CPSP. Administrators, teachers, students, and potentially parents could gain from the strategies and perspective that CPSP offers.

Since CPSP is a new framework, the body of scholarship on this topic should be supported by additional studies. As mentioned earlier in Chapter 3, it would be helpful to conduct similar studies that examine the intersections of science with gender, ability, or sexuality (Boda, 2019; Giordano, 2017). While critical pedagogy, culturally responsive, and multicultural frameworks have provided important insights regarding diversity in teaching diverse learners (Awad, 2011; Gay, 2000; Nieto & Bode, 2012), CPSP extends beyond this by using science to contextualize and critique social, cultural, historical, and contemporary processes toward deeper and more nuanced understandings of human diversity in society. Understanding the origins of the human family tree in Africa, and the development of different cultures and cultural narratives about the world in this way can help to decenter essentialist, racist, sexist, and/or religious fundamentalist discourse.

Another implication of the findings of the study is that CPSP could potentially be offered as a certificate program that is open to all majors. Such a program would provide in-depth training and credentialing documentation. Based on the findings of the study, the recommendations are aimed toward colleges of education, preservice, in-service teachers, and K-12 education.

Limitations

The study examined one case of three undergraduate PST education courses bounded to one institution, to determine the role of science in teaching about diversity in TEPs (Creswell, 1998). Given the inherent currency of this topic about which little is known, the findings may be applicable beyond the boundaries of this research. These limitations do not impinge on the quality of the study or its ability to respond to the research questions. Instead, by utilizing a single unit of analysis and small sample size, the design can be easily replicated and/or expanded. Moreover, this book could lay important groundwork for future research, including alternative applications of CPSP in professional development, curriculum development, and educational policy.

Future research should be conducted using CPSP in K-12 settings as well as in higher education courses (education and noneducation courses). In addition, further research should be conducted to determine the extent to which CPSP might inform administrative practices and educational policy.

Summary

In conclusion, this book investigated the role of science in teaching about human diversity, which places science at the center of inquiry. The research explored issues of science and diversity in a 21st-century context. Given the profound implications of increasing diversity in schools and the majority White teacher workforce, there is a growing need to prepare teachers to understand and teach about diversity. As such, this book is significant in that it explores the utility of science as a transdisciplinary method for teaching diversity to PSTs. CPSP has the potential for raising greater awareness about issues of race and ethnicity in schools and society. The findings of the study suggest that using science to discuss diversity can, indeed, influence positive and inclusive teacher practices that promote greater awareness and effective classroom practices in diverse settings.

One long-term implication of this approach goes beyond the classroom. Students who are taught to use a scientific lens to understand human diversity may also provide a basis from which to assuage fears of "the other," which underlie much of the current racial unrest. For example, the police involved killings of Black men and women generally involve an officer who believed that their life was in danger, despite having the only weapon in many of these tragic deaths. However, armed instead with knowledge of our common humanity, it is unlikely that the officers would use fear as justification for excessive force and/or killing someone with whom they cannot identify.

From a CPSP perspective, such fear might be viewed as socially learned behavior conveyed through ignorance of human similarity; that each person is valuable and should be treated as such. In this view, there is a direct line between social conditioning concerning race and the systemic, institutional inertia of racial discrimination. To disregard this connection serves only to perpetuate the notion of difference as an obstacle and inclusion as a chore. Were it the predominant view of society that race is not a biological principle but merely a stratifying feature, each individual's humanity would be acknowledged and afforded the necessary respect.

Since science is inextricably linked to our greater understandings of humanity, it is therefore imperative for PST programs to actively address these issues. CPSP offers science as a powerful tool with which to disentangle misconceptions about race in schools and society, and thus engender more meaningful conversations and mindful actions based on our common humanity.

References

Abrams, E., Taylor, P. C., & Guo, C.-J. (2013). Contextualizing culturally relevant science and mathematics teaching for indigenous learning. *International Journal of Science and Mathematics Education*, *11*(1), 1–21. https://doi.org/10.1007/s10763-012-9388-2

Adams, H. H. (1987). African and African-American contributions to science and technology. *Portland Baseline Essays*. Portland, OR: Portland Public Schools.

Agger, B. (1991). Critical theory, poststructuralism, postmodernism: Their sociological relevance. *Annual Review of Sociology*, *17*, 105–131.

Agin, D. (2006). *Junk Science: How politicians, corporations, and other hucksters betray us* (1st ed.). New York, NY: Thomas Dunne Books.

Aikenhead, G. S. (1996). Science education: Border crossing into the subculture of science. *Studies in Science Education*, *27*(1), 1–52. https://doi.org/10.1080/03057269608560077

Aikenhead, G. S. (1997). Student views on the influence of culture on science. *International Journal of Science Education, 19*(4), 419–428. https://doi.org/10.1080/0950069970190405

Akom, A. (2011). Eco-apartheid: Linking environmental health to educational outcomes. *Teacher's College Record*, *113*(4), 831–859.

Allen, C. D., & Eisenhart, M. (2017). Fighting for desired versions of a future self: How young women negotiated STEM-related identities in the discursive landscape of educational opportunity. *Mind, Culture, and Activity*, *26*(3), 407–436.

Altbach, P. G. (1971). Education and neocolonialism: A note. *Comparative Education Review*, *15*(2), 237–239.

Alway, J. (1995). *Critical theory and political possibilities: Conceptions of emancipatory politics in the works of Horkheimer, Adorno, Marcuse, and Habermas*. Westport, CT: Greenwood Press.

Amat, A. (2019). Tensions and transformations in a project based on agroecology. In L. A. Bryan & K. Tobin (Eds.), *Critical issues and bold visions for science education: The road ahead* (pp. 171–185). Leiden, The Netherlands: Brill.

American Anthropological Association. (1994). *American Anthropological Association statement on "race" and intelligence*. Retrieved from http://www.aaanet.org/stmts/racepp.htm

AAPA statement on biological aspects of race. (1996).*American Journal of Physical Anthropology*, *101*(4), 569–570. doi:10.1002/ajpa.1331010408

American Anthropological Association. (1998). *American Anthropological Association statement on "race."* Retrieved from http://www.aaanet.org/stmts/racepp.htm

American Association for the Advancement of Science. (1989). *Science for all Americans*. New York, NY: Oxford University Press.

American Association for the Advancement of Science. (2010). *Exploring the nature of science: Using the Atlas of Science Literacy and other education resources from AAAS Project 2061*. Washington, DC: Author.

Anderson, J. D. (1988). *The education of Blacks in the South, 1860–1935*. Chapel Hill, NC: University of North Carolina Press.

Apple, M. W. (1978). Ideology, reproduction, and educational reform. *Comparative Education Review*, *22*(3), 367–387.

Apple, M. W. (1993). The politics of official knowledge: Does a national curriculum make sense? *Discourse: Studies in the Cultural Politics of Education*, *14*(1), 1–16. https://doi.org/10.1080/0159630930140101

Apple, M. W. (2000). *Official knowledge: Democratic education in a conservative age* (2nd ed.). New York, NY: Routledge.

Apple, M. W. (2004). *Ideology and curriculum*. New York, NY: RoutledgeFalmer.

Apple, M. W. (2010). Theory, research, and the critical scholar/activist. *Educational Researcher*, *39*(2), 152–155. https://doi.org/10.3102/0013189X10362591

Apple, M. W. (2011a). The tasks of the critical scholar/activist in education: The contribution of José Gimeno Sacristán. *Revista de Educación*, *356*, 235–250.

Apple, M. W. (2011b). Global crises, social justice, and teacher education. *Journal of Teacher Education*, *62*(2), 222–234. https://doi.org/10.1177/0022487110385428

Archer, M. S. (1995). *Realist Social Theory: The morphogenetic approach*. Cambridge, UK: Cambridge University Press.

Areddy, J. T. (2020, May 26). China rules out animal market and lab as coronavirus origin. *The Wall Street Journal*. Retrieved June 13, 2020, from https://www.wsj.com/articles/china rules-out-animal-market-and-lab-as-coronavirus-origin-11590517508

Aristotle. (350 B.C.E./1966). *Metaphysics* (Vol. 1). (W.D. Ross, Trans.). Oxford, UK: Oxford Clarendon Press (Original work published in 350 B.C.E.).

Aronowitz, S. (1988). *Science as power: Discourse and ideology in modern society*. Minneapolis, MN: University of Minnesota Press.

Aronowitz, S., & Giroux, H. A. (1991). *Postmodern education: Politics, culture, and social criticism*. Minneapolis, MN: University of Minnesota Press.

Asante, M. K. (1990). *Kemet, Afrocentricity, and knowledge*. Trenton, NJ: Africa World Press.

Asante, M. K. (1991). The Afrocentric idea in education. *The Journal of Negro Education*, *60*(2), 170–180. https://doi.org/10.2307/2295608

Asante, M. K. (2000). *The Egyptian philosophers: Ancient African voices from Imhotep to Akhenaten*. Chicago, IL: African American Images.

Asante, M. K. (2010). Afrocentricity and Africology: Theory and practice in the discipline. In J.R. Davidson (Ed.) *African American studies* (pp. 35–52). Edinburgh, Scotland: Edinburgh University Press.

Asante, M. K., & Ravitch, D. (1991). Multiculturalism: An exchange. *The American Scholar*, *60*(2), 267–276.

Ash, A., & Wiggan, G. (2018). Race, multiculturalisms and the role of science in teaching diversity: Towards a critical postmodern science pedagogy. *Multicultural Education Review*, *10*(2), 94–120. https://doi.org/10.1080/2005615X.2018.1460894

Atwater, M. M. (1989). Including multicultural education in science education: Definitions, competencies, and activities. *Journal of Science Teacher Education*, *1*(1), 17–20. https://doi.org/10.1007/BF03032129

Atwater, M. M. (1996a). Teacher education and multicultural education: Implications for science education research. *Journal of Science Teacher Education*, 7(1), 1–21. https://doi.org/10.1007/BF00118343

Atwater, M. M. (1996b). Social constructivism: Infusion into the multicultural science education research agenda. *Journal of Research in Science Teaching*, *33*(8), 821–837. https://doi.org/10.1002/(SICI)1098-2736(199610)33:8<821::AID-TEA1>3.0.CO;2-Y.

Atwater, M. M. (2010). Multicultural science education and curriculum materials. *Science Activities: Classroom Projects and Curriculum Ideas*, *47*(4), 103–108. https://doi.org/10.1080/00368121003631652

Atwater, M. M. (2011). Significant science education research on multicultural science education, equity, and social justice. *Journal of Research in Science Teaching*, *49*(1), O1–O5. https://doi.org/10.1002/tea.20453.

Atwater, M. M., & Riley, J. P. (1993). Multicultural science education: Perspectives, definitions, and research agenda. *Science Education*, *77*(6), 661–668. https://doi.org/10.1002/sce.3730770609

Atwater, M. M., Russell, M., & Butler, M. B. (Eds.). (2013). *Multicultural science education: Preparing teachers for equity and social justice*. Dordrecht, The Netherlands: Springer Verlag.

Au, K. (1998). Social constructivism and the school literacy learning of students of diverse backgrounds. *Journal of Literacy Research*, *30*(2), 297–319. https://doi.org/10.1080/10862969809548000

Awad, I. (2011). Critical multiculturalism and deliberative democracy: Opening spaces for more inclusive communication. *Javnost – The Public*, *18*(3), 39–54.

Avraamidou, L. (2018). Stories we live, identities we build: How are elementary teachers' science identities shaped by their lived experiences. *Cultural Studies of Science Education*, *14*(2019), 33–59.

Balfanz, R., Byrnes, V., & Fox, J. (2012). *Sent home and put off-track: The antecedents, disproportionalities, and consequences of being suspended in the ninth grade*. Presented at the National Conference on Race and Gender Disparities in Discipline.

Ball, A. (2006). *Multicultural strategies for education and social change: Carriers of the torch in the United States and South Africa. Multicultural education series*. New York, NY: Teacher College Press.

Bandura, A. (1989). Human agency in social cognitive theory. *American Psychologist*, *44*(9), 1175–1184.

Bandura, A. (1999). Social cognitive theory: An agentic perspective. *Asian Journal of Social Psychology*, 2(1), 21–41. https://doi.org/10.1111/1467-839X.00024

Bandura, A. (2000). Exercise of human agency through collective efficacy. *Current Directions in Psychological Science*, 9(3), 75–78. https://doi.org/10.1111/1467-8721.00064

Bandura, A. (2001). Social cognitive theory: An agentic perspective. *Annual review of psychology*, *52*, 1–26. https://doi.org/10.1146/annurev.psych.52.1.1

Banks, J. A. (1993a). Multicultural education: Historical development, dimensions, and practice. *Review of Research in Education*, *19*, 3–49. https://doi.org/10.2307/1167339

Banks, J. A. (1993b). The canon debate, knowledge construction, and multicultural education. *Educational Researcher*, 22(5), 4–14. https://doi.org/10.3102/0013189X022005004

Banks, J. A. (2004). Teaching for social justice, diversity, and citizenship in a global world. *The Educational Forum*, *68*(4), 296–305. https://doi.org/10.1080/00131720408984645

Banks, J. A. (2006). Series forward. In A. F. Ball (Ed.), *Multicultural strategies for education and social change: Carriers of the torch in the United States and South Africa* (pp. xi–xv). New York, NY: Teacher College Press.

Banks, J. A., & Banks, C. A. M. (Eds.). (2004). *Handbook of research on multicultural education* (2nd ed.). San Francisco: Jossey-Bass.

Banks, J. A., & Banks, C. A. M. (2010). *Multicultural education: Issues and perspectives* (7th ed.). Hoboken, NJ: Wiley.

Barnes, C. J. (2006). Preparing preservice teachers to teach in a culturally responsive way. *Negro Educational Review*, *57*(1/2), 85–100.

Barton, A. C. (2003). *Teaching science for social justice*. New York, NY: Teachers College Press.

Barton, A. C., & Berchini, C. (2013). Becoming an insider: Teaching science in urban settings. *Theory into Practice*, *52*(1), 21–27. https://doi.org/10.1080/07351690.2013.743765

Barton, A. C., & McLaren, P. (2001). Capitalism, critical pedagogy, and urban science education: An interview with Peter McLaren. *Journal of Research in Science Teaching*, *38*(8), 847–859. https://doi.org/10.1002/tea.1035

Barton, A. C., & Yang, K. (2000). The culture of power and science education: Learning from Miguel. *Journal of Research in Science Teaching*, *37*(8), 871–889. https://doi.org/10.1002/1098-2736(200010)37:8<871::AID-TEA7>3.0.CO;2-9

Basu, S. J., & Barton, A. C. (2010). A researcher–student–teacher model for democratic science pedagogy: Connections to community, shared authority, and critical science agency. *Equity & Excellence in Education*, *43*(1), 72–87. https://doi.org/10.1080/10665680903489379.

Baudrillard, J. (1994). *Simulacra and simulation*. Ann Arbor, MI: University of Michigan.

Bauman, Z. (2000). *Liquid modernity*. Cambridge, UK: Polity Press.

Baxter, P., & Jack, S. (2008). Qualitative case study methodology: Study design and implementation for novice researchers. *The Qualitative Report*, *13*(4), 544–559.

Bazzul, J., & Siry, C. (Eds.). (2019). *Critical voices in science education research: Narratives of hope and struggle* (Vol. 17). Cham, Switzerland: Springer.

Beck, C. (1993). Postmodernism, pedagogy, and philosophy of education. *Philosophy of Education*, *27*, 1–13.

Bell, D. A. (1995a). Who's afraid of critical race theory. *University of Illinois Law Review*, *1995*, 893.

Bell, D. A. (1995b). Brown v. Board of Education and the interest- convergence dilemma. In K. Crenshaw, N. Gotanda, G. Peller, & K. Thomas (Eds.), *Critical race theory: The key writings that formed the movement* (pp. 20–29). New York, NY: New Press.

Ben-Jochannan, Y. (1973). *A chronology of the bible: A brief history of the development of the old and new testaments from their African and Asian origins to their European and European-American revisions, versions, etc.* Baltimore, MD: Black Classic Press.

Ben-Jochannan, Y. (1991). *African origins of the major "western religions."* Baltimore, MD: Black Classic Press.

Ben-Jochannan, Y. (1997). *Africa: Mother of western civilization.* Baltimore, MD: Black Classic Press.

Bennett, S. V. (2013). Effective facets of a field experience that contributed to eight preservice teachers' developing understandings about culturally responsive teaching. *Urban Education*, *48*(3), 380–419. https://doi.org/10.1177/0042085912452155.

Benton-Borghi, B. H., & Chang, Y. M. (2012). Critical examination of candidates' diversity competence: Rigorous and systematic assessment of candidates' efficacy to teach diverse student populations. *Teacher Educator*, 47(1), 29–44. https://doi.org/10.1080/08878730.2011.632472

Bernal, M. (1987). *Black Athena: The Afroasiatic roots of classical civilization.* New Brunswick, NJ: Rutgers University Press.

Bernal, M. (1996). [Book Review]. Mary Lefkowitz, Not out of Africa: How Afrocentrism became an excuse to teach myth as history. *Bryn Mawr Classical Review*. Retrieved from http://bmcr.brynmawr.edu/1996/96.04.05.html

Bernstein, R. (1983). *Beyond objectivism and relativism science, hermeneutics, and praxis*. Philadelphia, PA: University of Pennsylvania Press. Retrieved from http://ezproxy.viu.ca/login?url=http://muse.jhu.edu/books/9780812205503/

Bertoni, A., Goonan, K., Bonds, D., Whitt, M., Goff, D., & Brancati, F. (2005). Racial and ethnic disparities in cardiac catheterization for acute myocardial infarction in the United States, 1995—2001. *Journal of the National Medical Association*, 97(3), 317–323.

Best, S., & Kellner, D. (1991). *Postmodern theory: Critical interrogations*. New York, NY: Guilford Press.

Best, S., & Kellner, D. (2003). Contemporary youth and the postmodern adventure. *Review of Education, Pedagogy, and Cultural Studies*, *25*(2), 75–93. https://doi.org/10.1080/10714410390198949

Bianchini, J. A. (2012). Expanding underrepresented minority participation: America's science and technology talent at the crossroads. *Science Education*, 97(1), 163–166. https://doi.org/10.1002/sce.21032

Binder, A. J. (2000). Why do some curricular challenges work while others do not? The case of three Afrocentric challenges. *Sociology of Education*, *73*(2), 69–91.

Blake, N., & Masschelein, J. (2003). Critical theory and critical pedagogy. In N. Blake, P. Smeyers, R. Smith, & P. Standish (Eds.), *The Blackwell guide to the philosophy of education* (pp. 38–56). Malden, MA: Blackwell Publishing.

Blank, R. K. (2013). Science instructional time is declining in elementary schools: What are the implications for student achievement and closing the gap? *Science Education*, 97(6), 830–847.

Blumer, H. (1969). *Symbolic interactionism: Perspective and method*. Englewood Cliffs, NJ: Prentice Hall.

Boda, P. A. (2019). Conceptualizing the margins in science education: The limits of multicultural analyses. *Cultural Studies in Science Education*, *14*, 493–514.

Boeije, H. (2010). *Analysis in qualitative research*. Los Angeles, CA: Sage.

Bohman, J. (2004). Toward a critical theory of globalization: Democratic practice and multiperspectival inquiry. *Concepts and Transformation*, *9*(2), 121–146. https://doi.org/10.1075/cat.9.2.05boh

Bohman, J. (2005). We, heirs of enlightenment: Critical theory, democracy and social science. *International Journal of Philosophical Studies*, *13*(3), 353–377. https://doi.org/10.1080/09672550500169166

Bouchard, C. (1988). Genetic basis of racial differences. *Canadian Journal of Sport Sciences = Journal Canadien des Sciences du Sport*, *13*, 104.

Boulware, D. R., Pullen, M. F., Bangdiwala, A. S., Pastick, K. A., Lofgren, S. M., Okafor, E. C., Skipper, C. P., et al. (2020). A randomized trial of hydroxychloroquine as postexposure prophylaxis for COVID-19. *New England Journal of Medicine*, null. https://doi.org/10.1056/NEJMoa2016638

Bourdieu, P., & Passeron, J.-C. (1990). *Reproduction in education, society and culture* (R. Nice, Trans.). London: Sage.

Bowers, C. A. (2005). Is transformative learning the Trojan horse of western globalization? *Journal of Transformative Education*, *3*(2), 116–125. https://doi.org/10.1177/1541344604273622

Bowers, C. A. (2008). Why a critical pedagogy of place is an oxymoron. *Environmental Education Research*, *14*(3), 325–335.

Brandt-Rauf, P. W., & Brandt-Rauf, S. I. (1987). History of occupational medicine: Relevance of Imhotep and the Edwin Smith Papyrus. *British Journal of Industrial Medicine*, *44*(1), 68–70.

Braun, V., & Clarke, V. (2006). Using thematic analysis in psychology. *Qualitative Research in Psychology*, *3*(2), 77–101. https://doi.org/10.1191/1478088706qp063oa

Brew, G. (2019, May 23). The United States, OPEC, and international oil. *Oxford Research Encyclopedia of American History*. Retrieved June 13, 2020, from https://oxfordre.com/americanhistory/view/10.1093/acrefore/9780199329175.001.0001/crefore-9780199329175-e-719

Broadfoot, P. (2002). Structure and agency in education: The role of comparative education. *Comparative Education*, *38*(1), 5–6. https://doi.org/10.1080/03050060120103810

Brown, B. (2006). "It isn't no slang that can be said about this stuff": Language, identity, and appropriating science discourse. *Journal of Research in Science Teaching*, *43*(1), 96–126. https://doi.org/10.1002/tea.20096

Brown, M. (2007). Educating all students: Creating culturally responsive teachers, classrooms, and schools. *Intervention in School and Clinic*, *43*(1), 57–62. https://doi.org/10.1177/10534512070430010801

Bruna, K. R. (2016). A struggle for the soul: Reversing the odd alchemy of science schooling. *Mind, Culture, and Activity*, *23*(3), 259–269.

Bryan, L. A., & Tobin, K. (2019). *Critical Issues and Bold Visions for Science Education: The Road Ahead*. Leiden, The Netherlands: Brill.

Buczynski, S., & Hansen, C. B. (2010). Impact of professional development on teacher practice: Uncovering connections. *Teaching and Teacher Education*, *26*(3), 599–607. https://doi.org/10.1016/j.tate.2009.09.006

Burgess, D. J., Van Ryn, M., Crowley-Matoka, M., & Malat, J. (2006). Understanding the provider contribution to race/ethnicity disparities in pain treatment: Insights from dual process models of stereotyping. *Pain Medicine*, 7(2), 119–134. https://doi.org/10.1111/j.1526-4637.2006.00105.x

Buxton, C. A. (2010). Social problem solving through science: An approach to critical, place-based, science teaching and learning. *Equity & Excellence in Education*, *43*(1), 120–135. https://doi.org/10.1080/10665680903408932

Cable News Network (CNN). (2015, March 5). Oldest known jawbone from human genus found in Ethiopia. *News*. Retrieved from http://www.cnn.com/2015/03/05/africa/ethiopia-ancient-jawbone-discovery/

California Newsreel. (2003). *Race literacy quiz: What difference makes a difference?* Retrieved from http://newsreel.org/guides/race/quiz.htm

Campbell, G., Denes, R., & Morrison, C. (2000). *Access denied: Race, ethnicity, and the scientific enterprise*. Oxford: Oxford University.

Canella, G. S. (2007). Critical qualitative research. In G. Ritzer (Ed.), *The Blackwell encyclopedia of sociology* (pp. 867–870). Malden, MA: Blackwell Publishing.

Capobianco, B. M. (2007). Science teachers' attempts at integrating feminist pedagogy through collaborative action research. *Journal of Research in Science Teaching*, *44*(1), 1–32. https://doi.org/10.1002/tea.20120

Carnoy, M., & Rothstein, R. (2013). *What do international tests really show about U.S. student performance?*. Washington, DC: Economic Policy Institute. Retrieved from http://www.epi.org/files/2013/EPI-What-do-international-tests-really-show-about-US-student-performance.pdf

Carroll, J., & Minkler, M. (2000). Freire's message for social workers: Looking back, looking ahead. *Journal of Community Practice*, *8*(1), 21–36.

Carter, L. (2004). Thinking differently about cultural diversity: Using postcolonial theory to (re)read science education. *Science Education*, *88*(6), 819–836. https://doi.org/10.1002/sce.20000

Carter, L. (2006). Postcolonial interventions within science education: Using postcolonial ideas to reconsider cultural diversity scholarship. *Educational Philosophy and Theory*, *38*, 677–691. https://doi.org/10.1111/j.1469-5812.2006.00219.x

Carter, L. (2008). Globalization and science education: The implications of science in the new economy. *Journal of Research in Science Teaching*, *45*(5), 617–633. https://doi.org/10.1002/tea.20189

Carter, N., Larke, P. J., Singleton-Taylor, G., & Santos, E. (2003). Multicultural science education: Moving beyond tradition. In S. M. Hines (Ed.), *Multicultural science education: Theory, practice, and promise*. New York, NY: Peter Lang.

Castagno, A. E. (2009). Making sense of multicultural education: A synthesis of the various typologies found in the literature. *Multicultural Perspectives*, *11*(1), 43–48.

Castagno, A. E. (2013). Multicultural education and the protection of whiteness. *American Journal of Education*, *120*(1), 101–128.

Castro, A. J. (2010). Themes in the research on preservice teachers' views of cultural diversity implications for researching millennial preservice teachers. *Educational Researcher*, *39*(3), 198–210. https://doi.org/10.3102/0013189X10363819

Centers for Disease Control and Prevention. (2020a). Coronavirus (COVID-19). *Centers for Disease Control and Prevention*. Retrieved June 12, 2020, from https://www.cdc.gov/coronavirus/2019-ncov.

Centers for Disease Control and Prevention. (2020b, June). Cases in the U.S. *Centers for Disease Control and Prevention*. Retrieved June 12, 2020, from https://www.cdc.gov/coronavirus/2019-ncov/cases-updates/cases-in-us.html.

Center for Disease Control and Prevention. (2020, June 4). COVID-19 in racial and ethnic minority groups. *Center for Disease Control and Prevention*. Retrieved from https://www.cdc.gov/coronavirus/2019-ncov/need-extra-precautions/racial-ethnic-minorities.html.

Chakraborty, J., Collins, T., & Grineski, S. (2016). Environmental justice research: Contemporary issues and emerging topics. *International Journal of Environmental Research and Public Health*, *13*(11), 1072. https://doi.org/10.3390/ijerph13111072

Chen, E. W., & Yoo, G. J. (2010). *Encyclopedia of Asian American issues today*. Santa Barbara, CA: Greenwood Press.

Chen, J., Rathore, S. S., Radford, M. J., Wang, Y., & Krumholz, H. M. (2001). Racial differences in the use of cardiac catheterization after acute myocardial infarction. *New England Journal of Medicine*, *344*(19), 1443–1449. https://doi.org/10.1056/NEJM200105103441906

Church, J. A., & White, N. J. (2011). Sea-level rise from the late 19th to the early 21st century. *Surveys in Geophysics*, *32*(4–5), 585–602. https://doi.org/10.1007/s10712-011-9119-1

Chenoweth, K. (2007). *"It's being done": Academic success in unexpected schools*. Boston, MA: Harvard University Press.

Chenoweth, K. (2009). *How it's being done*. Cambridge, MA: Harvard University.

Chen, X. (2009). *Students who study science, technology, engineering, and mathematics (STEM) in postsecondary education* (Statistics in Brief No. NCES 2009-161) (pp. 1–24). Washington, DC: U.S. Department of Education.

Cho, G., & DeCastro-Ambrosetti, D. (2005). Is ignorance bliss? Pre-service teachers' attitudes toward multicultural education. *The High School Journal*, *89*(2), 24–28. https://doi.org/10.1353/hsj.2005.0020

Choi, J. (2008). Unlearning colorblind ideologies in education class. *Educational Foundations*, 22(3–4), 53–71.

Cibulka, J. G. (2014, May 23). *Letter from national council for accreditation of teacher education*. Retrieved from http://education.uncc.edu/sites/education.uncc.edu/files/media/Action%20Letter%20NCATE.pdf

Clarke, J. H. (1978). Ahmed Baba, a scholar of old Africa. *Negro History Bulletin*, *41*(6), 918–919.

Clarke, J. H. (1979). African-American historians and the reclaiming of African history. *Présence Africaine*, *110*(2), 29–48.

Cochran-Smith, M. (1995). Color blindness and basket making are not the answers: Confronting the dilemmas of race, culture, and language diversity in teacher education. *American Educational Research Journal*, *32*(3), 493–522. https://doi.org/10.3102/00028312032003493

Cochran-Smith, M. (2003). The multiple meanings of multicultural teacher education: A conceptual framework. *Teacher Education Quarterly*, *30*(2), 7–26.

Cochran-Smith, M. (2010). Toward a theory of teacher education for social justice. In A. Hargreaves, A. Lieberman, M. Fullan, & D. Hopkins (Eds.), *Second International Handbook of Educational Change*, Springer International Handbooks of Education (pp. 445–467). London: Springer.

Cochran-Smith, M., & Zeichner, K. M. (Eds.). (2009). *Studying teacher education: The report of the AERA Panel on research and teacher education*. Mahwah, NJ: Lawrence Erlbaum Associates.

Collins, P. H. (1998). It's all in the family: Intersections of gender, race, and nation. *Hypatia*, *13*(3), 62–82. https://doi.org/10.2307/3810699

Collins, P. H. (2000). *Black feminist thought: Knowledge, consciousness, and the politics of empowerment*. New York, NY: Routledge.

Combe, G. (1830). *A system of phrenology* (3rd ed.). Edinburgh, Scotland: J. Anderson.

Cook, J., Nuccitelli, D., Green, S. A., Richardson, M., Winkler, B., Painting, R., Way, R., et al. (2013). Quantifying the consensus on anthropogenic global warming in the scientific literature. *Environmental Research Letters*, *8*(2), 024024. https://doi.org/10.1088/1748-9326/8/2/024024

Cook, J., Oreskes, N., Doran, P. T., Anderegg, W. R. L., Verheggen, B., Maibach, E. W., Carlton, J. S., et al. (2016). Consensus on consensus: A synthesis of consensus estimates on human-caused global warming. *Environmental Research Letters*, *11*(4), 048002. https://doi.org/10.1088/1748-9326/11/4/048002

Correll, S. J. (2001). Gender and the career choice process: The role of biased self-assessments. *American Journal of Sociology*, *106*(6), 1691–1730. https://doi.org/10.1086/321299

Correa, N. (2016). *It's our community: The state of Jefferson County*. Southeast Texas Nonprofit Development Center. Retrieved from https://setxnonprofit.org/wp content/uploads/2016/03/Its-Our-Community.pdf

Corsiglia, J., & Snively, G. (2001). Rejoinder: Infusing indigenous science i nto western modern science for a sustainable future. *Science Education*, *85*(1), 82–86. https://doi.org/10.1002/1098-237X(200101)85:1<82::AID-SCE11>3.0.CO;2-Q

Crawford, B. (2014). Authentic science. In R. Gunstone (Ed.), *Encyclopedia of Science Education* (pp. 1–3). Netherlands: Springer.

Creswell, J. W. (1998). *Qualitative inquiry and research design: Choosing among five traditions*. Thousand Oaks, CA: Sage.

Creswell, J. W. (2005). *Educational research: Planning, conducting, and evaluating quantitative and qualitative research* (2nd ed.). Upper Saddle River, NJ: Merrill.

Creswell, J. W. (2007). *Qualitative inquiry and research design: Choosing among five traditions*. Thousand Oaks, CA: Sage.

Creswell, J. W. (2009). *Research design: Qualitative, quantitative, and mixed methods approaches* (3rd ed.). Los Angeles, CA: Sage.

Creswell, J. W. (2012). *Educational research: Planning, conducting, and evaluating quantitative and qualitative research* (4th ed.). Upper Saddle River, NJ: Pearson.

Creswell, J. W., & Miller, D. L. (2000). Determining validity in qualitative inquiry. *Theory into Practice*, *39*(3), 124–130. https://doi.org/10.1207/s15430421tip3903_2

Croizet, J.-C., & Dutrévis, M. (2004). Socioeconomic status and intelligence: Why test scores do not equal merit. *Journal of Poverty*, *8*(3), 91–107. https://doi.org/10.1300/J134v08n03_05

Cromley, J. G. (2009). Reading achievement and science proficiency: International comparisons from the Programme on International Student Assessment. *Reading Psychology*, *30*(2), 89–118. https://doi.org/10.1080/02702710802274903

Crotty, M. (1998). *The foundations of social research: Meaning and perspective in the research process*. Thousand Oaks, CA: Sage.

Cunningham, L. (2019). *Renewable energy and energy efficiency incentives: A summary of federal programs* (No. R40913) (p. 60). Washington, DC: Congressional Research Service. Retrieved from https://fas.org/sgp/crs/misc/R40913.pdf

Daley, B. J. (2004). Using concept maps in qualitative research. In *Proceedings of the first international conference on concept mapping* (pp. 14–17). Universidad Pública de Navarra, Pamplona, Spain.

Darling-Hammond, L. (2005). New standards and old inequalities: School reform and the education of African American students. In J. E. King (Ed.), *Black education: A transformative research and action agenda for the new century* (pp. 197–223). Mahwah, NJ: Lawrence Erlbaum Associates.

Darling-Hammond, L. (2010). *The flat world and education: How America's commitment to equity will determine our future*. Columbia, NY: Teachers College Press.

Delany-Barmann, G., & Minner, S. (1997). Development and implementation of a program of study to prepare teachers for diversity. *Equity & Excellence in Education*, *30*(2), 78–85. https://doi.org/10.1080/1066568970300209

De La Soul. (1996). *Stakes is high*. New York, NY: Tommy Boy.

Delgado, R. (1989). Storytelling for oppositionists and others: A plea for narrative. *Michigan Law Review*, *87*(8), 2411–2441.

Delgado, R., & Stefancic, J. (2001). Critical race theory: An introduction. *Critical America*. New York, NY: New York University Press.

Delpit, L. D. (2006). *Other people's children: Cultural conflict in the classroom*. New York, NY: New Press.

Delpit, L. D., & Dowdy, J. K. (2002). *The skin that we speak: Thoughts on language and culture in the classroom*. New York, NY: New Press.

deMarrais, K. B., & LeCompte, M. D. (1998). *The way schools work: A sociological analysis of education* (3rd ed.). White Plains, NY: Longman/Addison Wesley.

Denton, N., & Massey, D. S. (1993). *American apartheid: Segregation and the making of the underclass*. Cambridge, MA: Harvard University Press.

Denzin, N. K. (1986). Postmodern social theory. *Sociological Theory*, *4*, 194–204.

Denzin, N. K., & Lincoln, Y. S. (Eds.). (2005). *The Sage handbook of qualitative research* (3rd ed.). Thousand Oaks, CA: Sage Publications.

Derrida, J. (1978). *Writing and difference*. Chicago, IL: University of Chicago Press.

Derrida, J. (1982). *Positions*. (A. Bass, Trans.). Chicago, IL: University of Chicago Press.

Derrida, J. (2009). The decentering event in social thought. In C. C. Lemert (Ed.), *Social theory: The multicultural and classic readings* (pp. 413–417). Boulder, CO: Westview Press.

Devlin, B., Fienberg, S. E., Resnick, D. P., & Roeder, K. (Eds.). (1997). *Intelligence, genes, and success: Scientists respond to the bell curve*. New York, NY: Springer.

DeWalt, K. M., & DeWalt, B. R. (2011). *Participant observation: A guide for fieldworkers*. Lanham, MD: Rowman & Littlefield.

Dewey, J. (1916). *Democracy and education: An introduction to the philosophy of education*. Macmillan. Retrieved from http://www.gutenberg.org/ebooks/852

Dey, I. (1993). *Qualitative data analysis: A user-friendly guide for social scientists*. New York, NY: Routledge.

Diop, C. A. (1954). *Nations nègres et culture: de l'Antiquité nègre égyptienne aux problèmes culturels de l'Afrique noire d'aujourd'hui*. Paris: Présence Africaine.

Diop, C. A. (1974). *The African origin of civilization: Myth or reality*. New York, NY: Laurence Hill.

Diop, C. A. (1981). Origin of the Ancient Egyptians. In G. Mokhtar (Ed.), *General History of Africa II: Ancient Civilizations of Africa*, General History of Africa. London: Heinemann.

Dobson, P. J. (1999). Approaches to theory use in interpretive case studies: A critical realist perspective. Presented at the Australasian Conference on Information Systems, Wellington, New Zealand.

Donald, M. (1993). Précis of Origins of the modern mind: Three stages in the evolution of culture and cognition. *Behavioral and Brain Sciences*, *16*(4), 737–748. https://doi.org/10.1017/S0140525X00032647

Donovan, B. M., Semmens, R., Keck, P., Brimhall, E., Busch, K. C., Weindling, M., Duncan, A., Stuhlsatz, M., Bracey, Z., Bloom, M. Kowalski, S., & Salazar, B. (2019). Toward a more humane genetics education: Learning about the social and quantitative complexities of human genetic variation research could reduce racial bias in adolescent and adult populations. *Science Education*, *103*(3), 529–560. https://doi.org/10.1002/sce.21506

Du Bois, W. E. B. (1897). *The conservation of races*. The American Negro Academy Occasional Papers, No. 2. Washington, DC: American Negro Academy.

Du Bois, W. E. B. (1899). *The Philadelphia Negro: A social study*. Boston, MA: Ginn & Co.

Du Bois, W. E. B. (1903). *The souls of Black folk: Essays and sketches*. Chicago, IL: A. C. McClurg & Co.

Du Bois, W. E. B. (1911, August). Races. *The Crisis*, *2*(4), 157–159.

Duncan, A. (2017, March 2). Educational equality and excellence will drive a stronger economy. *The Brookings Institute*. Retrieved from https://www.brookings.edu/blog/brown-center-chalkboard/2017/03/02/educational-equality-and-excellence-will-drive-a-stronger-economy/

Durkheim, E. (2009a). Mechanic and organic solidarity. In C. C. Lemert (Ed.), *Social theory: The multicultural and classic readings* (pp. 73–77). Boulder, CO: Westview Press.

Durkheim, E. (2009b). Selected writings. In C. C. Lemert (Ed.), *Social theory: The multicultural and classic readings* (pp. 73–103). Boulder, CO: Westview Press.

Durodoye, B. A. (2003). The science of race in education. *Multicultural Perspectives*, *5*(2), 10–16.

Eagleton, T. (1991). *Ideology: An introduction*. New York, NY: Verso.

Eggers, M. L., & Massey, D. S. (1992). A longitudinal analysis of urban poverty: Blacks in U.S. metropolitan areas between 1970 and 1980. *Social Science Research*, *21*(2), 175–203. https://doi.org/10.1016/0049-089X(92)90014-8

Ehrenberg, R. G. (2010). Analyzing the factors that influence persistence rates in STEM field, majors: Introduction to the symposium. *Economics of Education Review*, *29*(6), 888–891. https://doi.org/10.1016/j.econedurev.2010.06.012

Eilks, I., Nielsen, J. A., & Hofstein, A. (2014). Learning about the role and function of science in public debate as an essential component of scientific literacy. In C. Bruguière, A. Tiberghien, & P. Clément (Eds.), *Topics and trends in current science education. Contributions from science education research* (pp. 85–100). Dordrecht, The Netherlands: Springer.

Einstein, A. (1920). *Relativity: The special and general theory*. New York, NY: Henry Holt.

Elias, S., & Feagin, J. R. (2016). *Racial theories in social science: A systemic racism critique*. New York, NY: Routledge.

Elmesky, R., & Tobin, K. (2005). Expanding our understandings of urban science education by expanding the roles of students as researchers. *Journal of Research in Science Teaching*, *42*(7), 807–828. https://doi.org/10.1002/tea.20079

Emdin, C. (2007). Exploring the contexts of urban science classrooms. Part 2: The emergence of rituals in the learning of science. *Cultural Studies of Science Education,* 2(2), 351–392. https://doi.org/10.1007/s11422-007-9057-x

Emdin, C. (2008a). Urban science classrooms and new possibilities: On intersubjectivity and grammar in the third space. *Cultural Studies of Science Education*, *4*(1), 239–254. https://doi.org/10.1007/s11422-008-9162-5

Emdin, C. (2008b). The three C's for urban science education. *Phi Delta Kappan*, *89*(10), 772–775.

Emdin, C. (2010). *Urban science education for the hip-hop generation*. Boston, MA: Sense.

Emdin, C. (2016). *For white folks who teach in the hood... and the rest of y'all too: Reality pedagogy and urban education*. Boston, MA: Beacon Press.

Environmental Protection Agency. (2015, December 23). Overview of Greenhouse Gases. *Environmental Protection Agency*. Overviews and Factsheets. Retrieved June 18, 2020, from https://www.epa.gov/ghgemissions/overview-greenhouse-gases.

Essink-Bot, M.-L., Lamkaddem, M., Jellema, P., Nielsen, S. S., & Stronks, K. (2013). Interpreting ethnic inequalities in healthcare consumption: A conceptual framework for research. *The European Journal of Public Health*, *23*(6), 922–926. https://doi.org/10.1093/eurpub/cks170

Feldman, S. (2001). Intersecting and contesting positions: Postcolonialism, feminism, and world-systems theory. *Review (Fernand Braudel Center)*, *24*(3), 343–371.

Fischer, C. S. (1996). *Inequality by design: Cracking the bell curve myth*. Princeton, NJ: Princeton University Press.

Fischman, G. E., & McLaren, P. (2005). Rethinking critical pedagogy and the Gramscian and Freirean legacies: From organic to committed intellectuals or critical pedagogy, commitment, and praxis. *Cultural Studies <=>Critical Methodologies,* *5*(4), 425–446. https://doi.org/10.1177/1532708605279701

Fleischman, L., & Franklin, M. (2017). *Fumes across the fence-line: The health impacts of air pollution from oil & gas facilities on African American communities*. NAACP. Retrieved from http://www.naacp.org/wp-content/uploads/2017/11/Fumes-Across-the-Fence Line_NAACP_CATF.pdf.

Foeman, A. K. (2009). Science and magic: DNA and the racial narratives that shape the social construction of race in the USA. *Intercultural Communication Studies*, *18*(2), 14–25.

Foucault, M. (1977). *Discipline and punish: The birth of the prison system* (A. Sheridan, Trans.). New York, NY: Vintage Books.

Foucault, M. (1980). *Power/knowledge: Selected interviews and other writings, 1972–1977*. New York, NY: Pantheon Books.

Foucault, M. (1997). Technologies of the self. In P. Rabinow (Ed.), *Ethics: Subjectivity and truth. Essential works of Foucault, 1954–1984* (Vol. 1, pp. 232–252). New York, NY: The New Press.

Foucault, M. (2002). *Archaeology of knowledge*. (A. Sheridan, Trans.) Routledge classics. New York, NY: Routledge.

Foucault, M. (2003). *Birth of the clinic*. London: Routledge.

Foucault, M. (2009). Power as knowledge. In C. C. Lemert (Ed.), *Social theory: The multicultural and classic readings* (pp. 473–479). Boulder, CO: Westview Press.

Fram, S. M. (2013). The constant comparative analysis method outside of grounded theory. *The Qualitative Report*, *18*(1), 1–25.

Freire, P. (1970). *Pedagogy of the oppressed*. New York, NY: Continuum.

Freire, P. (2005). *Teachers as cultural workers: Letters to those who dare teach*. Boulder, CO: Westview Press.

Freire, P., & Macedo, D. (1987). *Literacy: Reading the word & the world*. London: Routledge & Kegan Paul.

Funk, C., & Hefferon, M. (2019). U.S. Public views on climate change and energy. *Pew Research Center*. Retrieved from https://www.pewresearch.org/science/2019/11/25/u-s-public-views-on-climate-and-energy/.

Gall, J. P., Gall, M. D., & Borg, W. R. (2005). *Applying educational research: A practical guide* (5th ed.). Boston, MA: Pearson/Allyn & Bacon.

Gandin, L. A. (2011). Porto Alegre as a counter-hegemonic global city: Building globalization from below in governance and education. *Discourse: Studies in the Cultural Politics of Education*, *32*(2), 235–252. https://doi.org/10.1080/01596306.2011.562669

Garmon, M. A. (2004). Changing preservice teachers' attitudes/beliefs about diversity: What are the critical factors? *Journal of Teacher Education*, *55*(3), 201–213. https://doi.org/10.1177/0022487104263080

Garmon, M. A. (2005). Six key factors for changing preservice teachers' attitudes/beliefs about diversity. *Educational Studies*, *38*(3), 275–286. https://doi.org/10.1207/s15326993es3803_7

Gay, G. (2000). *Culturally responsive teaching: Theory, research, and practice*. Multicultural education series. New York: Teachers College Press.

Gay, G. (2002). Preparing for culturally responsive teaching. *Journal of Teacher Education*, *53*(2), 106–116. https://doi.org/10.1177/0022487102053002003

Gay, G. (2003). Developing cultural critical consciousness and self-reflection in preservice teacher education. *Theory into Practice*, *42*(3), 181–187. https://doi.org/10.1207/s15430421tip4203_3

Gay, G. (2010). Acting on beliefs in teacher education for cultural diversity. *Journal of Teacher Education*, *61*(1–2), 143–152. https://doi.org/10.1177/0022487109347320

Geuss, R. (1981). *The idea of a critical theory: Habermas and the Frankfurt School*. New York, NY: Cambridge University Press.

Gilbert, A. (2013). *Courageous pedagogy: Enacting critical science education*. Information Age Publishing.

Gill, D., & Levidow, L. (Eds.). (1987). *Anti-racist science teaching*. London: Free Association Books.

Giordano, S. (2017). Feminists increasing public understanding of science: A feminist approach to developing critical science literacy skills. *Frontiers: A Journal of Women Studies*, *38*(1), 100–123.

Giroux, H. A. (1988). *Teachers as intellectuals: Toward a critical pedagogy of learning*. Granby, MA: Bergin & Garvey.

Giroux, H. A. (1991). Border pedagogy and the politics of postmodernism. *Social Text*, 51–67.

Giroux, H. A. (2004). Critical pedagogy and the postmodern/modern divide: Towards a pedagogy of democratization. *Teacher Education Quarterly*, *31*(1), 31–47.

Giroux, H. A. (2005). *Border crossings: Cultural workers and the politics of education* (2nd ed.). New York, NY: Routledge.

Giroux, H. A. (2011). Fighting for the future: American youth and the global struggle for democracy. *Cultural Studies <=> Critical Methodologies*, *11*(4), 328–340. https://doi.org/10.1177/1532708611414658

Giroux, H. A., & Giroux, S. S. (2006). Challenging neoliberalism's new world order: The promise of critical pedagogy. *Cultural Studies <=> Critical Methodologies*, *6*(1), 21–32. https://doi.org/10.1177/1532708605282810

Glaser, B. G. (1978). *Theoretical sensitivity: Advances in the methodology of grounded theory*. Mill Valley, CA: Sociology Press.

Glaser, B. G., & Strauss, A. L. (1967). *The discovery of grounded theory: Strategies for qualitative research*. Chicago, IL: Aldine Publishing.

Glesne, C. (2011). *Becoming qualitative researchers: An introduction* (4th ed.). Boston, MA: Pearson.

Gordon, E. (2009). The global talent crisis. *The Futurist*, *43*(4), 34–39.

Gould, S. J. (1996). *The mismeasure of man*. New York, NY: W. W. Norton & Company.

Gramlich, J. (2020, May 6). Black imprisonment rate in the U.S. has fallen by a third since 2006. *Pew Research*. Retrieved June 12, 2020, from https://www.pewresearch.org/fact tank/2020/05/06/black-imprisonment-rate-in-the-u-s-has-fallen-by-a-third-since-2006/.

Gramsci, A. (1971). *Selections from the prison notebooks of Antonio Gramsci*. New York, NY: International Publishers.

Gravlee, C. C. (2009). How race becomes biology: Embodiment of social inequality. *American Journal of Physical Anthropology*, *139*, 47–57. https://doi.org/10.1002/ajpa.20983

Green, P. E. (1999). Separate and still unequal: Legal challenges to school tracking and ability grouping in America's public schools. In L. Parker, D. Deyhle, & S. A. Villenas (Eds.), *Race is... race isn't: Critical race theory and qualitative studies in education* (pp. 231–250). Boulder, CO: Westview Press.

Griffith, A. L. (2010). Persistence of women and minorities in STEM field majors: Is it the school that matters? *Economics of Education Review*, *29*(6), 911–922. https://doi.org/10.1016/j.econedurev.2010.06.010

Guba, E. G., & Lincoln, Y. S. (1994). Competing paradigms in qualitative research. In N. K. Denzin & Y. S. Lincoln (Eds.), *Handbook of qualitative research* (pp. 105–117). Thousand Oaks, CA: Sage.

Gunderson, R. (2014). *Nature, sociology, and The Frankfurt School* (Dissertation). East Lansing, MI: Michigan State University.

Gur-Ze'ev, I. (1998). Toward a nonrepressive critical pedagogy. *Educational Theory*, *48*(4), 463–486. https://doi.org/10.1111/j.1741-5446.1998.00463.x

Habermas, J. (1972). *Knowledge and human interests*. Boston, MA: Beacon Press.

Habermas, J. (1984). *The theory of communicative action* (Vol. 1). Boston, MA: Beacon Press.

Habermas, J. (1987). *The theory of communicative action* (Vol. 2). Boston, MA: Beacon Press.

Hall, S. (1992). Race, culture, and communications: Looking backward and forward at cultural studies. *Rethinking Marxism*, *5*(1), 10–18. doi:10.1080/08935699208657998

Hall, S. (1999). Encoding/decoding. In S. During (Ed.), *The cultural studies reader* (2nd ed., pp. 507–517). New York, NY: Routledge.

Hamilton, J. D. (1983). Oil and the macroeconomy since World War II. *Journal of Political Economy*, *91*(2), 228–248. https://doi.org/10.1086/261140

Hancock, D. R., & Algozzine, R. (2006). *Doing case study research: A practical guide for beginning researchers*. New York, NY: Teachers College Press.

Haraway, D. (1988). Situated knowledges: The science question in feminism and the privilege of partial perspective. *Feminist studies*, *14*, 575–599.

Haraway, D. (1991). *Simians, cyborgs and women: The reinvention of nature*. New York, NY: Routledge.

Haraway, D. (2006). A cyborg manifesto: Science, technology, and socialist-feminism in the late 20th century. *The international handbook of virtual learning environments* (pp. 117–158). Dordrecht, The Netherlands: Springer.

Harris, M. (1999). *Theories of culture in postmodern times*. Walnut Creek, CA: AltaMira Press.

Hartley, J. (2004). Case study research. In C. Cassell & G. Symon (Eds.), *Essential guide to qualitative methods in organizational research* (pp. 323–333). Thousand Oaks, CA: Sage.

Hartsock, N. (1989). Postmodernism and political change: Issues for feminist theory. *Cultural Critique*, 15–33.

Harvey, D. (2005). *A brief history of neoliberalism*. Oxford, UK: Oxford University Press.

hayes, kecia. (2006). Bring in da noise, bring in Du Bois: Infusing an African-American educational ideology into the urban education discourse. In J. L. Kincheloe, Kecia Hayes, K. Rose, & P. M. Anderson (Eds.), *The Praeger handbook of urban education* (pp. 77–89). Westport, CT: Greenwood Publishing Group.

Held, D. (1980). *Introduction to critical theory: Horkheimer to Habermas*. Polity Press.

Herodotus. (440 B.C.E./2014). *The history of Herodotus, volume 1*. (G.C. Macaulay, Trans.). Lexington, KY: CreateSpace Independent Publishing Platform. (Original work published in 440 B.C.E.).

Herrnstein, R., & Murray, C. (1994). *The bell curve: Intelligence and class structure in American life*. New York, NY: Free Press.

Hickling-Hudson, A. (2003). Multicultural education and the postcolonial turn. *Policy Futures in Education*, *1*, 381–401.

Hickling-Hudson, A. (2004). Educating teachers for cultural diversity and social justice. In G. Hernes (Ed.), *Planning for diversity: Education in multi-ethnic and multicultural societies* (pp. 270–307). Paris: International Institute for Education Planning (UNESCO).

Hickman, H. & Porfilio, B. J. (2012). *Problematizing the portrayal of marginalized groups in textbooks*. Rotterdam, The Netherlands: Sense Publishers.

Hilberth, M., & Slate, J. R. (2014). Middle school black and white student assignment to disciplinary consequences: A clear lack of equity. *Education and Urban Society*, *46*(3), 312–328. doi:10.1177/0013124512446218

Hilliard, A. G. (1986). Pedagogy in ancient Kemet. *Kemet and the African worldview: Research, rescue, and restorations* (pp. 138–139).

Hilliard, A. G. (1992). The meaning of KMT (Ancient Egyptian) history for contemporary African American experience. *Phylon (1960-)*, *49*, 10–22.

Hilliard, A. G. (2000). Excellence in education versus high-stakes standardized testing. *Journal of Teacher Education*, *51*(4), 293–304. https://doi.org/10.1177/0022487100051004005

Hilliard, A. G. (2012). Standards: Decoy or quality control? In W. Au & M. B. Tempel (Eds.), *Pencils down: Rethinking high-stakes testing and accountability in public schools*, A Rethinking Schools publication (pp. 16–20). Milwaukee, WI: Rethinking Schools.

Hilliard, A. G., Williams, L., & Damali, N. (1987). *The teachings of Ptahhotep: The oldest book in the world*. Atlanta, GA: Blackwood Press.

Himmelstein, J. L. (2014). How the mass media use numbers to tell a story: The case of the crack scare of 1986. *Advancing Education in Quantitative Literacy*, *7*(1), 1–24.

Hochman, G., Rajagopal, D., & Zilberman, D. (2010). The effect of biofuels on crude oil markets. *AgBioForum*, *13*(2), 112–118.

Hodson, D. (1993). In search of a rationale for multicultural science education. *Science Education*, *77*(6), 685–711. https://doi.org/10.1002/sce.3730770611

Hodson, D. (1999). Critical multiculturalism in science and technology education. In S. May (Ed.), *Critical multiculturalism: Rethinking multicultural and antiracist education* (pp. 236–266). Philadelphia, PA: Falmer Press.

Hodson, D., & Dennick, R. (1994). Antiracist education: A special role for the history of science and technology. *School Science and Mathematics*, *94*(5), 255–262. https://doi.org/10.1111/j.1949-8594.1994.tb15666.x

Hoepner, C. C. (2010). *Advanced placement math and science courses: Influential factors and predictors for success in college STEM majors* (Doctoral Dissertation). UCLA, Los Angeles, CA. Retrieved from ProQuest Digital Dissertations (AAT 3437520).

Honneth, A. (1979). Communication and reconciliation: Habermas' critique of Adorno. *Telos*, *1979*(39), 45–61. https://doi.org/10.3817/0379039045

hooks, bell. (1990). Postmodern blackness. *Postmodern Culture, 1*(1). https://doi.org/10.1353/pmc.1990.0004

hooks, bell. (1994). *Teaching to transgress: Education as the practice of freedom*. New York, NY: Routledge.

hooks, bell. (1999). A revolution of values: The promise of multicultural change. In S. During (Ed.), *The cultural studies reader* (2nd ed., pp. 233–267). New York, NY: Routledge.

Höök, M., & Tang, X. (2013). Depletion of fossil fuels and anthropogenic climate change—A review. *Energy Policy*, *52*, 797–809. https://doi.org/10.1016/j.enpol.2012.10.046

Hornik, K., Cutts, B., & Greenlee, A. (2016). Community theories of change: Linking environmental justice to sustainability through stakeholder perceptions in Milwaukee (WI, USA). *International Journal of Environmental Research and Public Health*, *13*(10), 979. https://doi.org/10.3390/ijerph13100979

Hopkins, T. K., & Wallerstein, I. M. (1982). *World-systems analysis: Theory and methodology*. Sage Publications.

Horkheimer, M. (1982). *Critical theory: Selected essays*. New York, NY: Continuum.

Horkheimer, M. (1993). *Between philosophy and social science: Selected early writings. Studies in contemporary German social thought*. Cambridge, MA: MIT Press.

Horkheimer, M., & Adorno, T. W. (1972). *Dialectic of enlightenment*. New York, NY: Seabury Press.

Howard, G. R. (2006). *We can't teach what we don't know: White teachers, multiracial schools. Multicultural education series* (2nd ed.). New York, NY: Teachers College Press.

Huang, C. (2011, June 6). Facebook and Twitter key to Arab Spring uprisings: Report. *The National: UAE*. Retrieved from http://www.thenational.ae/news/uae-news/facebook-and-twitter-key-to-arab-spring-uprisings-report

Huerta, A. (2020, May 15). The right to ethnic studies in higher education. *Inside Higher Ed*. Retrieved from https://www.insidehighered.com/advice/2020/05/15/why-students-should-be-required-take-ethnic-studies-opinion

Human Rights Watch. (2020, May 12). Covid-19 fueling anti-Asian racism and xenophobia worldwide. *Human Rights Watch*. Retrieved from https://www.hrw.org/news/2020/05/12/covid-19-fueling-anti-asian-racism-and-xenophobia-worldwide.

Hume, A., & Berry, A. (2011). Constructing CoRes—A strategy for building PCK in pre-service science teacher education. *Research in Science Education, 41*(3), 341–355. https://doi.org/10.1007/s11165-010-9168-3

Humes, K. R., Jones, N. A., & Ramirez, R. R. (2011). *Overview of race and Hispanic origin: 2010*. U.S. Government Printing Office.

Hurry, J. B. (1926). *Imhotep: The vizier and physician of King Zoser and afterwards the Egyptian god of medicine*. Oxford, UK: Oxford University Press, H. Milford.

Hurtado, S., Cabrera, N. L., Lin, M. H., Arellano, L., & Espinosa, L. L. (2008). Diversifying science: Underrepresented student experiences in structured research programs. *Research in Higher Education*, *50*(2), 189–214. https://doi.org/10.1007/s11162-008-9114-7

Hurtado, S., Newman, C. B., Tran, M. C., & Chang, M. J. (2010). Improving the rate of success for underrepresented racial minorities in STEM fields: Insights from a national project. *New Directions for Institutional Research*, *2010*(148), 5–15. https://doi.org/10.1002/ir.357

Hutchison, C. B. (2006). Cultural constructivism: The confluence of cognition, knowledge creation, multiculturalism, and teaching. *Intercultural Education*, *17*(3), 301–310. https://doi.org/10.1080/14675980600841694

Hutchison, C. B. (2009). *What happens when students are in the minority: Experiences that impact human performance*. Lanham, MD: Rowman & Littlefield.

Israel, S. (2017, November 1). Building America's future: STEM education intervention is a win-win. *Public Policy Initiative*. University of Pennsylvania's Wharton School of Public Policy. Retrieved from https://publicpolicy.wharton.upenn.edu/live/news/2188-building-americas-future-stem-education

Irvine, J. J., & York, D. E. (1993). Teacher perspectives: Why do African American, Hispanic, and Vietnamese students fail? In S. W. Rothstein (Ed.), *Handbook of schooling in urban America* (pp. 161–173). Westport, CT: Greenwood Press.

Irvine, J. J. (1992). Making teacher education culturally responsive. In M. E. Dilworth (Ed.), *Diversity in teacher education: New expectations* (pp. 79–82). San Francisco: Jossey-Bass.

Jacobs, A., Richtel, M., & Baker, M. (2020, March 19). 'At war with no ammo': Doctors say shortage of protective gear is dire. *The New York Times*. Retrieved from https://www.nytimes.com/2020/03/19/health/coronavirus-masks-shortage.html

James, G. G. M. (2010). *Stolen legacy: The Egyptian origins of western philosophy*. Almanor, CA: Feather Trail Press.

Jameson, F. (1998). *The cultural turn: Selected writings on the postmodern, 1983–1998*. Verso Books.

Jegede, O. J., & Aikenhead, G. S. (1999). Transcending cultural borders: Implications for science teaching. *Research in Science & Technological Education*, *17*(1), 45–66. https://doi.org/10.1080/0263514990170104

Johnson, A., Brown, J., Carlone, H., & Cuevas, A. K. (2011). Authoring identity amidst the treacherous terrain of science: A multiracial feminist examination of the journeys of three women of color in science. *Journal of Research in Science Teaching*, *48*(4), 339–366. https://doi.org/10.1002/tea.20411

Johnson, C. C., Kahle, J. B., & Fargo, J. D. (2007). Effective teaching results in increased science achievement for all students. *Science Education*, *91*(3), 371–383. https://doi.org/10.1002/sce.20195

Johnson, C. C., & Marx, S. (2009). Transformative professional development: A model for urban science education reform. *Journal of Science Teacher Education, 20*(2), 113–134. https://doi.org/10.1007/s10972-009-9127-x

Jones, C. F. (2010). A landscape of energy abundance: Anthracite coal canals and the roots of American fossil fuel dependence, 1820–1860. *Environmental History*, *15*(3), 449–484. https://doi.org/10.1093/envhis/emq057

Jost, J. T., Babera, P., Bonneau, R., Langer, M., Metzger, M., Nagler, J., Sterling, J., Tucker, J. A. (2018). How social media facilitates political protest: Information, motivation, and social networks. *Advances in Political Psychology*, *39*(1), 85–118.

Kahn, R., & Kellner, D. (2003). Internet subcultures and oppositional politics. In D. Muggleton & R. Weinzierl (Eds.), *The post-subcultures reader* (pp. 299–313). New York, NY: Berg.

Kamarck, E. (2019). The challenging politics of climate change. *Brookings Institute*. Retrieved from https://www.brookings.edu/research/the-challenging-politics-of-climate-change/.

Kamentez, A. (2016, April 12). Pearson's quest to cover the planet in company-run schools. *Wired*. Retrieved from https://www.wired.com/2016/04/apec-schools/.

Kanpol, B. (1992). *Towards a theory and practice of teacher cultural politics: Continuing the postmodern debate*. Norwood, NJ: Ablex.

Kanpol, B., & McLaren, P. (Eds.). (1995a). Introduction: Resistance multiculturalism and the politics of difference. *Critical multiculturalism: Uncommon voices in a common struggle* (pp. 1–17). Westport, CT: Bergin & Garvey.

Kanpol, B., & McLaren, P. (Eds.). (1995b). *Critical multiculturalism: Uncommon voices in a common struggle*. Westport, CT: Bergin & Garvey.

Kaplan, G. A., Pamuk, E. R., Lynch, J. W., Cohen, R. D., & Balfour, J. L. (1996). Inequality in income and mortality in the United States: Analysis of mortality and potential pathways. *BMJ*, *312*(7037), 999–1003. https://doi.org/10.1136/bmj.312.7037.999

Kawachi, I., & Kennedy, B. P. (1999). Income inequality and health: Pathways and mechanisms. *Health Services Research*, *34*(1 Pt 2), 215–227.

Kayumova, S., Avraamidou, L., & Adams, J. D. (2019). Diversity, equity and the big picture. In L. A. Bryan & K. Tobin (Eds.), *Critical issues and bold visions for science education: The road ahead* (pp. 285–297). Leiden, The Netherlands: Brill.

Kellner, D. (1990). Critical theory and the crisis of social theory. *Sociological Perspectives, 33*(1), 11–33. https://doi.org/10.2307/1388975

Kellner, D. (2000). Multiple literacies and critical pedagogies. In P. P. Trifonas (Ed.), *Revolutionary pedagogies cultural politics, instituting education, and the discourse of theory*. New York, NY: Routledge.

Kellner, D. (2003). Globalization, technopolitics, and revolution. In J. Foran (Ed.), *The future of revolutions: Rethinking political and social change in the age of globalization* (pp. 180–193). London: Zed Books.

Kellner, D. (2009). Dialectics of globalization: From theory to practice. In S. Dasgupta & J. Nederveen Pieterse (Eds.), *Politics of globalization* (pp. 179–196). Thousand Oaks, CA: Sage.

Kennedy, B. P., Kawachi, I., & Prothrow-Stith, D. (1996). Income distribution and mortality: Cross sectional ecological study of the Robin Hood index in the United States. *BMJ*, *312*(7037), 1004–1007. https://doi.org/10.1136/bmj.312.7037.1004

Kilduff, M., & Mehra, A. (1997). Postmodernism and organizational research. *The Academy of Management Review*, *22*(2), 453–481. https://doi.org/10.2307/259330

Kilgore, D. (2004). Toward a postmodern pedagogy. *New Directions for Adult and Continuing Education*, *2004*(102), 45–53. https://doi.org/10.1002/ace.137

Kincheloe, J. L. (2001). Describing the bricolage: Conceptualizing a new rigor in qualitative research. *Qualitative Inquiry*, *7*(6), 679–692. https://doi.org/10.1177/107780040100700601

Kincheloe, J. L. (2004). *Critical pedagogy primer*. New York, NY: Peter Lang.

Kincheloe, J. L. (2005). *Critical constructivism primer*. New York, NY: Peter Lang.

Kincheloe, J. L. (2008). *Knowledge and critical pedagogy: An introduction*. New York, NY: Springer.

Kincheloe, J. L. (2010). Why a book on urban education? In S. R. Steinberg (Ed.), *19 urban questions: Teaching in the city* (2nd ed., pp. 1–25). New York, NY: Peter Lang.

Kincheloe, J. L. (2011a). Critical pedagogy and the knowledge wars of the twenty-first century. In Kecia Hayes, S. R. Steinberg, & K. Tobin (Eds.), *Key works in critical pedagogy* (pp. 385–405). Boston, MA: Sense Publishers.

Kincheloe, J. L. (2011b). Meet me behind the curtain: The struggle for a critical postmodern action research. In Kecia Hayes, S. R. Steinberg, K. Tobin, & J. L. Kincheloe (Eds.), *Key works in critical pedagogy*, Bold Visions in Educational Research (Vol. 32, pp. 85–99). Rotterdam, The Netherlands: Sense Publishers.

Kincheloe, J. L., & Berry, K. (2004). *Rigour and complexity in educational research: Conceptualizing the bricolage*. London: Open University Press.

King, J. E. (1991). Dysconscious racism: Ideology, identity, and the miseducation of teachers. *The Journal of Negro Education*, *60*(2), 133. https://doi.org/10.2307/2295605

Kirmaci, M., Buxton, C. A., & Allexsaht-Snider, M. (2019). "Being on the other side of the table": A qualitative study of a community-based science learning program with Latinx families. *Urban Education*. https://doi.org/10.1177/0042085919877934

Kozol, J. (2005). *The shame of the nation: The restoration of apartheid schooling in America*. New York, NY: Crown Publishers.

Kuhn, T. S. (1962). *The structure of scientific revolutions*. Chicago, IL: University of Chicago Press.

Kumar, R., & Hamer, L. (2012). Preservice teachers' attitudes and beliefs toward student diversity and proposed instructional practices: A sequential design study. *Journal of Teacher Education*, *64*(2), 162–177. https://doi.org/10.1177/0022487112466899

Kumashiro, K. K. (2001). "Posts" perspectives on anti-oppressive education in social studies, English, mathematics, and science classrooms. *Educational Researcher*, *30*(3), 3–12. https://doi.org/10.3102/0013189X030003003

Kumashiro, K. K. (2008). *The seduction of common sense: How the right has framed the debate on America's schools*. New York, NY: Teachers College Press.

Kunjufu, J. (2002). *Black students/middle class teachers*. Chicago, IL: African American Images.

Kvale, S. (2007). *Doing interviews*. Thousand Oaks, CA: Sage.

Kyne, D., & Bolin, B. (2016). Emerging environmental justice issues in nuclear power and radioactive contamination. *International Journal of Environmental Research and Public Health*, *13*(7), 700. https://doi.org/10.3390/ijerph13070700

Ladislaw, S., & Barnett, J. (2019). *Oil and natural gas development*. Energy in America (pp. 4–10). Center for Strategic and International Studies (CSIS). Retrieved from www.jstor.org/stable/resrep22550.5

Ladson-Billings, G. (1995a). But that's just good teaching! The case for culturally relevant pedagogy. *Theory into Practice*, *34*(3), 159–165. https://doi.org/10.1080/00405849509543675

Ladson-Billings, G. (1995b). Toward a theory of culturally relevant pedagogy. *American Educational Research Journal*, *32*(3), 465–491. https://doi.org/10.3102/00028312032003465

Ladson-Billings, G. (1998). Just what is critical race theory and what's it doing in a nice field like education? *International Journal of Qualitative Studies in Education*, *11*(1), 7–24. https://doi.org/10.1080/095183998236863

Ladson-Billings, G. (2000). Fighting for our lives: Preparing teachers to teach African American students. *Journal of Teacher Education*, *51*(3), 206–214. https://doi.org/10.1177/0022487100051003008

Ladson-Billings, G. (2006). From the achievement gap to the education debt: Understanding achievement in U.S. schools. *Educational Researcher*, *35*(7), 3–12. https://doi.org/10.3102/0013189X035007003

Ladson-Billings, G. (2009). *The dreamkeepers: Successful teachers of African American children*. San Francisco, CA: Jossey-Bass Publishers.

Larson, E. J. (1995). *Sex, race, and science: Eugenics in the Deep South*. Baltimore, MD: Johns Hopkins University Press.

Lather, P. (1992). Critical frames in educational research: Feminist and post-structural perspectives. *Theory into Practice*, *31*(2), 87–99. https://doi.org/10.1080/00405849209543529

Laughter, J. C., & Adams, A. D. (2012). Culturally relevant science teaching in middle school. *Urban Education*, *47*(6), 1106–1134. https://doi.org/10.1177/0042085912454443

Law, J. (2008). On sociology and STS. *The Sociological Review*, *56*(4), 623–649.

Le, P. T., & Matias, C. E. (2018). Towards a truer multicultural science education: How whiteness impacts science education. *Cultural Studies in Science Education*, *14*(2019), 15–31.

Leake, D. O., & Leake, B. L. (1992). African-American immersion schools in Milwaukee: From the inside. *Phi Delta Kappa*, *73*(10), 783–785.

LeCompte, M. D. (2000). Analyzing qualitative data. *Theory into Practice*, *39*(3), 146–154. https://doi.org/10.1207/s15430421tip3903_5

Lee, C. D. (2008). Synthesis of research on the role of culture in learning among African American youth: The contributions of Asa G. Hilliard, III. *Review of Educational Research*, *78*(4), 797–827. https://doi.org/10.3102/0034654308320967

Lee, J., & Bean, F. D. (2010). *The diversity paradox: Immigration and the color line in twenty-first century America*. New York, NY: Russell Sage Foundation.

Lee, O. (2001). Culture and language in science education: What do we know and what do we need to know? *Journal of Research in Science Teaching*, *38*(5), 499–501. https://doi.org/10.1002/tea.1015

Lee, O. (2011). Effective STEM education strategies for diverse and underserved learners. Presented at the Workshop on Successful STEM Education in K-12 Schools. Washington, DC: National Academies Press. Retrieved from http://sites.nationalacademies.org/

Lee, O., & Buxton, C. (2008). Science curriculum and student diversity: A framework for equitable learning opportunities. *The Elementary School Journal*, *109*(2), 123–137. https://doi.org/10.1086/590522

Lee, O., & Fradd, S. H. (1998). Science for all, including students from non-English-language backgrounds. *Educational Researcher*, 27(4), 12–21. https://doi.org/10.3102/0013189X027004012

Lee, V., Robinson, S., & Sebastian, J. (2012). The quality of instruction in urban high schools: Comparing mathematics and science to English and social studies classes in Chicago. *The High School Journal*, *95*(3), 14–48. https://doi.org/10.1353/hsj.2012.0006

Lefkowitz, M. R. (1997). *Not out of Africa: How Afrocentrism became an excuse to teach myth as history*. New York, NY: Basic Books.

Lemert, C. C. (Ed.). (2009). *Social theory: The multicultural and classic readings*. Boulder, CO: Westview Press.

Lincoln, Y. S., & Guba, E. G. (1985). *Naturalistic inquiry*. Newbury Park, CA: Sage.

Lipman, P. (2000). Bush's education plan, globalization, and the politics of race. *Cultural Logic, 4*(1). Retrieved from http://clogic.eserver.org/4-1/lipman.html

Lipman, P. (2002). Making the global city, making inequality: The political economy and cultural politics of Chicago school policy. *American Educational Research Journal, 39*(2), 379–419. https://doi.org/10.3102/00028312039002379

Lipman, P. (2004). *High stakes education: Inequality, globalization, and urban school reform*. New York, NY: RoutledgeFalmer.

Lorde, A. (1984). *Sister outsider: Essays and speeches*. Berkeley, CA: Crossing Press.

Loving, C. C. (1997). From the summit of truth to its slippery slopes: Science education's journey through positivist-postmodern territory. *American Educational Research Journal, 34*(3), 421–452. https://doi.org/10.3102/00028312034003421

Lowell, B. L., & Martin, P. (2012). Managing the dynamic science and engineering labor market in the United States. *International Migration Review, 46*(4), 1005–1012. https://doi.org/10.1111/imre.12008

Luck, L., Jackson, D., & Usher, K. (2006). Case study: A bridge across the paradigms. *Nursing Inquiry, 13*(2), 103–109. https://doi.org/10.1111/j.1440-1800.2006.00309.x

Lumpkin, B. (1984). The pyramids: Ancient showcase of African science and technology. *Journal of African Civilizations, 5*(1–2), 67–83.

Lyiscott, J. (2014, February). *3 ways to speak English* [Video file]. Retrieved from https://www.ted.com/talks/jamila_lyiscott_3_ways_to_speak_english.

Lyotard, J.-F. (1984). *The postmodern condition: A report on knowledge*. Manchester, UK: Manchester University Press.

Lyotard, J.-F. (2009). The postmodern condition. In C. C. Lemert (Ed.), *Social theory: The multicultural and classic readings* (pp. 465–468). Boulder, CO: Westview Press.

Macdonald, W. L. (1998). *English speaking migrant children in educational and cultural transition* (Dissertation). Curtin University of Technology, Perth, Australia. Retrieved from http://espace.library.curtin.edu.au/view/action/nmets.do?DOCCHOICE=10503.xml&dvs=1426756312151~779&locale=en_US&search_terms=&adjacency=&VIEWER_URL=/view/action/nmets.do?&DELIVERY_RULE_ID=4&divType=

Mackie, C. (2020, April 21). STEM and coronavirus: The crisis within the crisis. *Forbes*. Retrieved from https://www.forbes.com/sites/calvinmackie/2020/04/21/stem-and-coronavirus-the-crisis-within-the-crisis/#673003d6e662

MacMillan, A. (2016, March 11). Global warming 101. *NRDC*. Retrieved June 12, 2020, from https://www.nrdc.org/stories/global-warming-101

Majno, G. (1975). *The healing hand: Man and wound in the ancient world*. Cambridge, MA: Harvard University Press.

Maltby, C. (2003). *Journey of man: A genetic odyssey*. [The story of the human species - Hosted by Dr. Spencer Wells]. PBS Video.

Mannheim, K. (1954). *Ideology and utopia: An introduction to the sociology of knowledge*. New York, NY: Harcourt, Brace & Co. Retrieved from http://archive.org/details/ideologyutopiain00mann

Mansour, N., & Wegerif, R. (2013a). Why science education for diversity? In N. Mansour & R. Wegerif (Eds.), *Science education for diversity: Theory and practice* (pp. ix–xx). New York, NY: Springer.
Mansour, N., & Wegerif, R. (2013b). *Science education for diversity: Theory and practice*. New York, NY: Springer.
Marcuse, H. (2001). *Towards a critical theory of society*. (D. Kellner, Ed.) (Vol. 2). New York, NY: Routledge.
Marcuse, H. (2009). *Negations: Essays in critical theory*. (J. J. Shapiro, Trans.). London: MayFly Books.
Marks, J. (1995). Human biodiversity: Genes, race, and history. *Foundations of human behavior*. New York, NY: Aldine de Gruyter
Marks, J. (1996). Science and race. *American Behavioral Scientist*, *40*(2), 123–133. https://doi.org/10.1177/0002764296040002003
Marks, J. (2008). Scientific racism, History of. (J. H. Moore, Ed.) *Encyclopedia of race and racism, Macmillan social science library*. Detroit, MI: Macmillan Reference USA.
Marshall, C., & Rossman, G. B. (2006). *Designing qualitative research*. Thousand Oaks, CA: Sage.
Marshack, A. (1972). *The roots of civilisation*. London: McGraw Hill. Harvard University Press.
Marx, K. (2009a). Selected writings. In C. C. Lemert (Ed.), *Social theory: The multicultural and classic readings* (pp. 32–67). Boulder, CO: Westview Press.
Marx, K. (2009b). Labour-power and capital. In C. C. Lemert (Ed.), *Social theory: The multicultural and classic readings* (pp. 62–67). Boulder, CO: Westview Press.
Massey, D. S. (2004). Segregation and stratification: A biosocial perspective. *Du Bois Review: Social Science Research on Race*, *1*(01), 7–25. https://doi.org/10.1017/S1742058X04040032
Massey, D. S., & Eggers, M. L. (1990). The ecology of inequality: Minorities and the concentration of poverty, 1970–1980. *American Journal of Sociology*, *95*, 1153–1188.
Matthews, M. R. (2008). Science, worldviews and education: An introduction. *Science & Education*, *18*(6–7), 641–666. https://doi.org/10.1007/s11191-008-9170-6
May, S. (Ed.). (1999). *Critical multiculturalism: Rethinking multicultural and antiracist education*. Philadelphia, PA: Falmer Press.
Metcalf, G. (2020, June). What should carbon cost? *Scientific American*, *322*(6), 63–69.
McCarthy, N. (2017, February 2). The countries with the most STEM graduates. *Forbes*. Retrieved from https://www.forbes.com/sites/niallmccarthy/2017/02/02/the-countries-with-the-most-stem-graduates-infographic/#6f5363f3268a
McCarthy, T. A. (1985). *The critical theory of Jürgen Habermas*. Cambridge, MA: MIT Press.
McComas, W. F. (2013). *The language of science education: An expanded glossary of key terms and concepts in science teaching and learning*. Rotterdam, The Netherlands: Sense Publishers.
McComas, W. F. (2014). Pedagogical content knowledge (PCK). In W. F. McComas (Ed.), *The language of science education* (pp. 71–71). Boston, MA: Sense Publishers.

McConnell, T. J., Parker, J. M., & Eberhardt, J. (2013). Assessing teachers' science content knowledge: A strategy for assessing depth of understanding. *Journal of Science Teacher Education*, *24*(4), 717–743.

McFalls, E. L., & Cobb-Roberts, D. (2001). Reducing resistance to diversity through cognitive dissonance instruction: Implications for teacher education. *Journal of Teacher Education*, *52*(2), 164–172. https://doi.org/10.1177/0022487101052002007

McIntosh, P. (1989). White privilege: Unpacking the invisible knapsack. *Peace and Freedom*, 10–12.

McKnight, J. (1997). *Straight science? Homosexuality, evolution and adaptation*. New York, NY: Routledge.

McLaren, P. (1993). Multiculturalism and the postmodern critique: Towards a pedagogy of resistance and transformation. *Cultural Studies*, *7*(1), 118–146. https://doi.org/10.1080/09502389300490101

McLaren, P. (1999). *Schooling as a ritual performance: Toward a political economy of educational symbols and gestures* (3rd ed.). Lanham, MD: Rowman & Littlefield.

McLaren, P. (2001). Ché Guevara, Paulo Freire, and the politics of hope: Reclaiming critical pedagogy. *Cultural Studies <=> Critical Methodologies, 1*(1), 108–131.

McLaren, P. (2007). *Life in schools: An introduction to critical pedagogy in the foundations of education* (5th ed.). Boston, MA: Pearson/Allyn and Bacon.

McLaren, P., & Farahmandpur, R. (2001). Teaching against globalization and the new imperialism: Toward a revolutionary pedagogy. *Journal of Teacher Education*, *52*(2), 136–150. https://doi.org/10.1177/0022487101052002005

McLaren, P., Martin, G., Farahmandpur, R., & Jaramillo, N. (2004). Teaching in and against the empire: Critical pedagogy as revolutionary praxis. *Teacher Education Quarterly, 31*(1), 131–154.

McLaren, P., & Torres, R. D. (1999). Racism and multicultural education: Rethinking "race" and "whiteness" in late capitalism. In S. May (Ed.), *Critical multiculturalism: Rethinking multicultural and antiracist education* (pp. 42–76). Philadelphia, PA: Falmer Press.

Mensah, F. M. (2011). A case for culturally relevant teaching in science education and lessons learned for teacher education. *The Journal of Negro Education*, *80*(3), 296–309.

Mensah, F. M., & Jackson, I. (2018). Whiteness as property in science teacher education. *Teacher College Record*, *120*, 1–38.

Miles, M. B., & Huberman, A. M. (1994). *Qualitative data analysis: An expanded sourcebook*. Thousand Oaks, CA: Sage.

Miller, P. M., Brown, T., & Hopson, R. (2011). Centering love, hope, trust in the community: Transformative urban leadership informed by Paulo Freire. *Urban Education*, *46(5)*, 1078–1099.

Milner, H. R. (2006). Preservice teachers' learning about cultural and racial diversity: Implications for urban education. *Urban Education*, *41*(4), 343–375. https://doi.org/10.1177/0042085906289709

Milner, H. R. (2012). Rethinking achievement gap talk in urban education. *Urban Education*, *48*(1), 3–8. https://doi.org/10.1177/0042085912470417

Mohai, P., Pellow, D., & Roberts, J. T. (2009). Environmental justice. *Annual Review of Environment and Resources*, *34*(1), 405–430. https://doi.org/10.1146/annurev-environ-082508-094348

Moosa-Mitha, M. (2005). Situating anti-oppressive theories within critical and difference centred perspectives. In L. A. Brown & S. Strega (Eds.), *Research as resistance: Critical, indigenous and anti-oppressive approaches* (pp. 37–72). Toronto, ON, Canada: Canadian Scholars Press.

Moraes, M., & McLaren, P. (2003). The path of dissent: An interview with Peter McLaren. *Journal of Transformative Education, 1*(2), 117–134. https://doi.org/10.1177/1541344603001002004

Morrison, K. (1995). *Marx, Durkheim, Weber: Formations of modern social thought*. Thousand Oaks, CA: Sage.

Moscovici, H. (2008). Science teacher retention in today's urban schools: A study of success and failure. *Urban Education*, *44*(1), 88–105. https://doi.org/10.1177/0042085908318527

Mouffe, C., & Holdengräber, P. (1989). Radical democracy: Modern or postmodern? *Social text*, 31–45.

Moustakas, C. E. (1994). *Phenomenological research methods*. Thousand Oaks, CA: Sage.

Mukhopadhyay, C., & Henze, R. C. (2003). How real is race? Using anthropology to make sense of human diversity. *Phi Delta Kappan*, *84*(9), 669–678.

Mullis, I. V. S., Martin, M. O., Foy, P., & Stanco, G. M. (2012). *TIMSS 2011: International results in science*. Chestnut Hill, MA: TIMSS & PIRLS International Study Center.

Murfin, B. (1992, March). Science in school curriculum. Paper presented at the National Science Teachers Association (NSTA) National Convention. Boston, MA. Retrieved from https://www.africa.upenn.edu/K-12/African_Science.html

NASA. (2019, December 10). Greenland's rapid melt will mean more flooding. *NASA: Climate change and global warming*. Retrieved June 12, 2020, from https://climate.nasa.gov/news/2940/greenlands-rapid-melt-will-mean-more-flooding/

NASA. (2020a, June 11). Global climate change: Vital signs of the planet. *NASA: Climate change and global warming*. Retrieved June 12, 2020, from https://climate.nasa.gov/

NASA. (2020b, February 6). Arctic ice melt is changing ocean currents. *NASA: Climate change and global warming*. Retrieved June 12, 2020, from https://climate.nasa.gov/news/2950/arctic-ice-melt-is-changing-ocean-currents/

National Oceanic and Atmospheric Administration. (2018, June 25). What percentage of the American population lives near the coast? *National Ocean Service*. Retrieved June 18, 2020, from https://oceanservice.noaa.gov/facts/population.html

Nascimento, E. L. (2007). *The sorcery of color: Identity, race, and gender in Brazil*. Philadelphia, PA: Temple University Press.

National Assessment of Educational Progress (NAEP). (2015a). National achievement-level results. *Nation's Report Card*. Retrieved from https://www.nationsreportcard.gov/mathematics/nation/achievement/?grade=4

National Assessment of Educational Progress (NAEP). (2015b). National achievement-level results. *Nation's Report Card*. Retrieved from https://www.nationsreportcard.gov/mathematics/nation/achievement/?grade=8

National Assessment of Educational Progress (NAEP). (2019a). National achievement-level results. *Nation's Report Card*. Retrieved from https://www.nationsreportcard.gov/science_2015/#state/acl/chart_loc_2?grade=4

National Assessment of Educational Progress (NAEP). (2019b). National achievement-level results. *Nation's Report Card*. Retrieved from https://www.nationsreportcard.gov/science_2015/#state/acl/chart_loc_2?grade=8

National Association for Research in Science Teaching (NARST). (2018). *Creating a multicultural learning environment in science classrooms*. Retrieved from https://narst.org/research-matters/multicultural-learning-environment-science-classrooms

National Association for Multicultural Education (NAME). (2019). I teach science. Can I be a multicultural educator? *The National Association for Multicultural Education: Advancing and Advocating for Social Justice and Equity*. Retrieved from https://www.nameorg.org/learn/i_teach_science_can_i_be_a_mu.php

National Center for Education Statistics. (2017). *Number and percentage distribution of teachers in public and private elementary and secondary schools, by selected teacher characteristics: Selected years, 1987–88 through 2015–16*. U.S. Department of Education. Retrieved from https://nces.ed.gov/programs/digest/d17/tables/dt17_209.10.asp?current=yes

National Center for Education Statistics. (2018). *The condition of education 2018: Science performance*. U.S. Department of Education. Retrieved from https://nces.ed.gov/pubs2018/2018144.pdf

National Center for Education Statistics. (2019a). *The condition of education 2019: Education expenditures by country*. Washington, DC: U.S. Department of Education. Retrieved from http://nces.ed.gov/programs/digest/d13/tables/dt13_605.10.asp

National Center for Education Statistics. (2019b). *The condition of education 2019* (No. NCES 2019-144). Washington, DC: U.S. Department of Education. Retrieved from https://nces.ed.gov/pubsearch/pubsinfo.asp?pubid=2019144

National Council for Accreditation of Teacher Education. (2011). *Unit Standards in effect 2008, Standard 4: Diversity*. Retrieved from http://www.ncate.org/Standards/UnitStandards/UnitStandardsinEffect2008/tabid/476/Default.aspx#stnd4

National Human Genome Research Institute. (2006, 2011). *Whole genome association studies*. Retrieved August 11, 2013, from http://www.genome.gov/17516714

National Library of Medicine. (2002, September 16). *Greek medicine*. Retrieved from https://www.nlm.nih.gov/hmd/greek/greek_oath.html.

National Research Council. (1996). *National science education standards*. Washington, DC: National Academies Press.

National Research Council. (2007a). *Rising above the gathering storm: Energizing and employing America for a brighter economic future*. Washington, DC: National Academies Press.

National Research Council. (2007b). *Beyond bias and barriers: Fulfilling the potential of women in academic science and engineering*. Washington, DC: National Academies Press.

National Research Council. (2007c). *Taking science to school: Learning and teaching science in grades K-8*. Washington, DC: National Academies Press.

National Research Council. (2010). *Rising above the gathering storm, revisited: Rapidly approaching category 5*. Washington, DC: National Academies Press.

National Research Council. (2011). *Expanding underrepresented minority participation: America's science and technology talent at the crossroads*. Washington, DC: National Academies Press.

National Research Council. (2012). *A framework for K-12 science education: Practices, crosscutting concepts, and core ideas*. Washington, DC: National Academies Press.

National Science Board. (2007). *A national action plan for addressing the critical needs of the U.S. Science, technology, engineering, and mathematics education system* (NSB-07-114). Retrieved from http://www.nsf.gov

National Science Board. (2010). *Preparing the next generation of STEM innovators: Identifying and developing our nation's human capital* (NSB-10-33). Retrieved from http://www.nsf.gov

National Science Foundation. (2020a). The state of U.S. science and engineering 2020: Science and engineering indicators. U.S. and Global Education: Degrees Awarded. *National Science Board*. Retrieved from https://ncses.nsf.gov/pubs/nsb20201/u-s-and-global-education

National Science Foundation. (2020b). The state of U.S. science and engineering 2020: Science and engineering indicators. U.S. Science and Engineering Workforce. *National Science Board*. Retrieved from https://ncses.nsf.gov/pubs/nsb20201/u-s-s-e-workforce

National Science Teachers Association. (2020). *Multicultural science education*. Retrieved from https://www.nsta.org/nstas-official-positions/multicultural-science-education

Nelson, A. (2002). Unequal treatment: Confronting racial and ethnic disparities in health care. *Journal of the National Medical Association*, *94*(8), 666–668.

Nembhard, J. G. (2005). On the road to democratic economic participation: Educating African American youth in the postindustrial global economy. In J. E. King (Ed.), *Black education a transformative research and action agenda for the new century* (pp. 225–239). Mahwah, NJ: Lawrence Erlbaum Associates.

Nesta, L., Vona, F., & Nicolli, F. (2014). Environmental policies, competition and innovation in renewable energy. *Journal of Environmental Economics and Management*, *67*(3), 396–411. https://doi.org/10.1016/j.jeem.2014.01.001

Newman, S. (2001). Derrida's deconstruction of authority. *Philosophy & Social Criticism*, 27(3), 1–20. https://doi.org/10.1177/019145370102700301

New York State School Boards Association. (2017). When politics enters the classroom: Teaching climate change in schools. *New York State School Boards Association*. Retrieved from http://www.nyssba.org/clientuploads/nyssba_pdf/report-teaching-climate-change-08312017.pdf

Next Generation Science Standards. (2013). Topic Arrangements for the Next Generation of Science Standard. *Next Generation Science Standards: For States, By States*. Retrieved from https://www.nextgenscience.org/sites/default/files/AllTopic.pdf

Nicol, B. (2009). *The Cambridge introduction to postmodern fiction* (Vol. 67). New York, NY: Cambridge University Press.

Nieto, S., & Bode, P. (2012). *Affirming diversity: The sociopolitical context of multicultural education* (6th ed.). Boston, MA: Prentice Hall.

Noble, D. F. (1993). *A world without women: The Christian clerical culture of Western science*. New York, NY: Oxford University Press.

Noblit, G. W. (2013). Culture bound: Science, teaching and research. *Journal of Research in Science Teaching*, *50*(2), 238–249. https://doi.org/10.1002/tea.21072

Nola, R., & Irzık, G. (2005). Philosophy, science, education and culture. *Science & technology education library*. Dordrecht, The Netherlands: Springer.

North Carolina Department of Public Instruction. (2019, December 10). NC remains a leader in the numbers of national board certified teachers. *North Carolina Department of Public Instruction*. Raleigh, NC. Retrieved from https://www.dpi.nc.gov/news/press-releases/2019/12/10/nc-remains-leader-numbers-national-board-certified-teachers

Noyce, R. N., & Hoff, M. E. (1981). A history of microprocessor development at Intel. *IEEE Micro, 1*(1), 8–21. https://doi.org/10.1109/MM.1981.290812

Nunez, C. (2019, January 22). Causes and effects of climate change. *National Geographic*. Retrieved June 12, 2020, from https://www.nationalgeographic.com/environment/global warming/global-warming-overview/

Oakes, J. (2005). *Keeping track: How schools structure inequality*. New Haven, CT: Yale University Press.

O'Connor, M. K., Netting, F. E., & Thomas, M. L. (2008). Grounded theory: Managing the challenge for those facing institutional review board oversight. *Qualitative Inquiry*, *14*(1), 28–45. https://doi.org/10.1177/1077800407308907

Omi, M., & Winant, H. (1994). *Racial formation in the United States: From the 1960s to the 1990s*. New York, NY: Routledge.

Ong, M., Wright, C. C., & Orfield, G. (2011). Inside the double-blind: A synthesis of empirical research on undergraduate and graduate women of color in science, technology, engineering, and mathematics. *Harvard Educational Review*, *81*(2), 172–209.

Ordover, N. (2003). *American eugenics: Race, queer anatomy, and the science of nationalism*. Minneapolis, MN: University of Minnesota.

Organisation for Economic Co-operation and Development. (2011a). Lessons from PISA for the United States. *Strong performers and successful reformers in education*. Paris: OECD Publishing.

Organisation for Economic Co-operation and Development. (2011b). *PISA 2009 results* (Vol. 1). Paris: OECD Publishing.

Organisation for Economic Co-operation and Development. (2013a). *PISA results in focus: What 15-year-olds know and what they can do with what they know*. Author. Retrieved from http://www.oecd.org/pisa/keyfindings/pisa-2012-results-overview.pdf

Organisation for Economic Co-operation and Development. (2013b). Education at a glance 2013: OECD indicators. *Education at a Glance*. Paris: OECD Publishing. Retrieved from http://www.oecd-ilibrary.org/education/education-at-a-glance-2013_eag-2013-en

Organisation for Economic Co-operation and Development. (2014a). *PISA 2012 results: What students know and can do* (Vol. 1). Paris: OECD Publishing.

Organisation for Economic Co-operation and Development. (2014b). Education at a glance 2014: Highlights. *Education at a Glance*. Paris: OECD Publishing.

Orlowski, P. (2011). *Teaching about hegemony race, class and democracy in the 21st century*. New York, NY: Springer.

Osborne, J., Simon, S., & Collins, S. (2003). Attitudes towards science: A review of the literature and its implications. *International Journal of Science Education*, *25*(9), 1049–1079. https://doi.org/10.1080/0950069032000032199

Osler, W. (1921). *The evolution of modern medicine*. New Haven, CT: Yale University Press.

Owen, A. D. (2004). Environmental externalities, market distortions and the economics of renewable energy technologies. *The Energy Journal, 25*(3). https://doi.org/10.5547/ISSN0195-6574-EJ-Vol25-No3-7

Park, S., Jang, J.-Y., Chen, Y.-C., & Jung, J. (2011). Is pedagogical content knowledge (PCK) necessary for reformed science teaching?: Evidence from an empirical study. *Research in Science Education*, *41*(2), 245–260. https://doi.org/10.1007/s11165-009-9163-8

Parsons, E. C., & Carlone, H. B. (2013). Culture and science education in the 21st century: Extending and making the cultural box more inclusive. *Journal of Research in Science Teaching*, *50*(1), 1–11. https://doi.org/10.1002/tea.21068

Patton, M. Q. (1990). *Qualitative evaluation and research methods* (2nd ed.). Newbury Park, CA: Sage.

Peshkin, A. (1988). In search of subjectivity. One's own. *Educational Researcher, 17*(7), 17–21. https://doi.org/10.2307/1174381

Pew Research Center. (2018, July 11). Activism in the social media age. *Pew Research Center.* Retrieved from https://www.pewresearch.org/internet/2018/07/11/activism-in-the-social-media-age/

Pharmacy and Medicine in Ancient Egypt. (n.d.). *University of Manchester's KNH Centre for Biomedical Egyptology*. Retrieved from http://www.knhcentre.manchester.ac.uk/research/previousandcompletedresearch/pharmacyproject

Plato. (348 B.C.E./1968). *Laws*. (R.G. Bury, Trans.). Cambridge, MA: Harvard University Press. (Original work published in 348 B.C.E.)

Pleasants, N. (2002). *Wittgenstein and the idea of a critical social theory: A critique of Giddens, Habermas and Bhaskar*. Routledge.

Plutarch. (75 C.E./2012). *Parallel lives vol. 1*. Radford, VA: SMK Books. (Original work published in 75 C.E.).

Polkinghorne, D. E. (2003). Narrative configuration in qualitative analysis. In J. A. Hatch & R. Wisniewski (Eds.), *Life history and narrative*, Qualitative studies series (pp. 5–23). Washington, DC: Falmer Press.

Pollock, M. (2006). Everyday antiracism in education. *Anthropology News*, *47*(2), 9–10.

Popkewitz, T. S. (1998). Dewey, Vygotsky, and the social administration of the individual: Constructivist pedagogy as systems of ideas in historical spaces. *American Educational Research Journal*, *35*(4), 535–570. https://doi.org/10.3102/00028312035004535

Popper, K. R. (1963). *Conjectures and refutations: The growth of scientific knowledge*. London: Routledge & Kegan Paul.

Popper, K. R. (1966). *The open society and its enemies* (5th ed., Vol. 1 & 2). Princeton, NJ: Princeton University Press.

Popper, K. R. (1996). *The myth of the framework: In defence of science and rationality*. Psychology Press.

Popper, K. R. (2002). *The logic of scientific discovery*. New York, NY: Routledge. Retrieved from http://site.ebrary.com/id/5006148

Portland Baseline Essays. (1987). *Portland Public Schools*. Retrieved from http://www.pps.k12.or.us/files/curriculum/prefc-af.pdf

Price, J. (2010). The effect of instructor race and gender on student persistence in STEM fields. *Economics of Education Review*, 29(6), 901–910.

Pride, L. D. (2019). Having mindful discussions about race, privilege, and social justice. In L. A. Bryan & K. Tobin (Eds.), *Critical issues and bold visions for science education: The road ahead* (pp. 127–55). Leiden, The Netherlands: Brill.

Priest, S. (2014). Critical science literacy: What citizens and journalists need to know to make sense of science. *Bulletin of Science, Technology and Society, 33*(5–6), 138–45.

Proctor, R. (1988). *Racial hygiene: Medicine under the Nazis*. Cambridge, MA: Harvard University.

Program for International Student Assessment (PISA). (2015). PISA 2015 results. *Students Well Being: Volume III* [Overview]. Retrieved from https://www.oecd.org/pisa/PISA-2015-Results-Students-Well-being-Volume-III-Overview.pdf

Program for International Student Assessment (PISA). (2018a). PISA 2018 results. *Organisation for Economic Co-operation and Development* (OECD). https://www.oecd.org/pisa/PISA-results_ENGLISH.png

Program for International Student Assessment (PISA). (2018b). PISA 2018: Insights and interpretations. *Organisation for Economic Co-operation and Development* (OECD). Retrieved from https://www.oecd.org/pisa/PISA%202018%20Insights%20and%20Interpretations%20FINAL%20PDF.pdf

Quigley, C. (2009). Globalization and science education: The implications for indigenous knowledge systems. *International Education Studies*, 2(1). https://doi.org/10.5539/ies.v2n1p76

Rahm, J., & Brandt, C. B. (2016). Reimagining science education in neoliberal global contexts: Sociocultural accounts of science learning in underserved communities. *Mind, Culture, and Activity, 23*(3), 183–187.

Raphelson, S. (2017, October 31). Flint residents confront long-term health issues after lead exposure. *National Public Radio*. Retrieved from https://www.npr.org/2017/10/31/561155244/flint-residents-confront-long-term-health-issues-after-lead-exposure

Raymond, O. U., Kuyoro, S. O., Adekunle, Y. A., & Awodele, O., (2013). Application of microprocessors. *International Journal of Emerging Technology and Advanced Engineering, 3*(4), 488–493. https://doi.org/10.1.1.413.7858

Rebell, M., & Wolff, J. (2008). *Moving every child ahead: From NCLB hype to meaningful educational opportunity*. New York, NY: Teachers College Press.

Reidmiller, D., Avery, C., Easterling, D., Kunkel, K., Lewis, K., Maycock, T., & Stewart, B. (2018). *Impacts, risks, and adaptation in the United States: Fourth national climate assessment, Volume II*. National Climate Assessment (p. 1515). Washington, DC: U.S. Global Research Program. Retrieved from https://doi.org/10.7930/NCA4.2018

Riegle-Crumb, C., & King, B. (2010). Questioning a white male advantage in STEM: Examining disparities in college major by gender and race/ethnicity. *Educational Researcher, 39*(9), 656–664. https://doi.org/10.3102/0013189X10391657

Riegle-Crumb, C., Moore, C., & Ramos-Wada, A. (2011). Who wants to have a career in science or math? Exploring adolescents' future aspirations by gender and race/ethnicity. *Science Education, 95*(3), 458–476. https://doi.org/10.1002/sce.20431

Ritt, M. (2016). *The impact of high-stakes testing on the learning environment*. (Dissertation). St. Catherine's University, St. Paul, MN.

Rivera Maulucci, M. S. (2013). Emotions and positional identity in becoming a social justice science teacher: Nicole's story. *Journal of Research in Science Teaching*, *50*(4), 453–478. https://doi.org/10.1002/tea.21081

Rodney, W. (2011). *How Europe underdeveloped Africa*. Baltimore, MD: Black Classic Press.

Rodriguez, A. J. (1998). Strategies for counterresistance: Toward sociotransformative constructivism and learning to teach science for diversity and for understanding. *Journal of Research in Science Teaching*, 35, 589–622. doi:10.1002/(SICI)1098-2736(199808)35:6<589::AID-TEA2>3.0.CO;2-I

Rodriguez, A. J. (2015). What about a dimension of engagement, equity, and diversity practices? A critique of the next generation science standards. *Journal of Research in Science Teaching*, *52*(7), 1031–1051.

Rodriguez, A. J., & Morrison, D. (2019). Expanding and enacting transformative meanings of equity, diversity and social justice in science education. *Cultural Studies of Science Education*, *14*, 265–281.

Rømer, T. A. (2011). Postmodern education and the concept of power. *Educational Philosophy and Theory*, *43*(7), 755–772. https://doi.org/10.1111/j.1469-5812.2009.00566.x

Rosario, V. A. (1997). Homosexual bio-histories: Genetic nostalgias and the quest for paternity. In V. A. Rosario (Ed.), *Science and homosexualities* (pp. 1–25). New York, NY: Routledge.

Ross, D. D., & Smith, W. (1992). Understanding preservice teachers' perspectives on diversity. *Journal of Teacher Education*, *43*(2), 94–103. https://doi.org/10.1177/002248719204300203

Rossman, G. B., & Rallis, S. F. (2003). *Learning in the field: An introduction to qualitative research* (2nd ed.). Thousand Oaks, CA: Sage.

Roth, W. M., & Lee, S. (2004). Science education as/for participation in the community. *Science Education*, *88*, 263–291.

Rowley, J. (2002). Using case studies in research. *Management Research News*, *25*(1), 16–27. https://doi.org/10.1108/01409170210782990

Rubin, H. J., & Rubin, I. (2005). *Qualitative interviewing: The art of hearing data* (2nd ed.). Thousand Oaks, CA: Sage.

Sandelowski, M. (1995). Qualitative analysis: What it is and how to begin. *Research in Nursing & Health*, *18*(4), 371–375. https://doi.org/10.1002/nur.4770180411

Sanders, A. R. D., Sandvik, P. T., & Storli, E. (2019). *The political economy of resource regulation: An international and comparative history, 1850–2015*. Retrieved from https://www.deslibris.ca/ID/457599

Schneider, R. M., & Plasman, K. (2011). Science teacher learning progressions: A review of science teachers' pedagogical content knowledge development. *Review of Educational Research*, *81*(4), 530–565. https://doi.org/10.3102/0034654311423382

Schulman, K. A., Berlin, J. A., Harless, W., Kerner, J. F., Sistrunk, S., Gersh, B. J., Dubé, R., et al. (1999). The effect of race and sex on physicians' recommendations for cardiac catheterization. *New England Journal of Medicine*, *340*(8), 618–626. https://doi.org/10.1056/NEJM199902253400806

Schott Foundation. (2012). "*The urgency of now: The Schott 50 state report on public education and black males*." Cambridge, MA. Retrieved from http://www.schottfoundation.org/urgency-of-now.pdf

Seale, C. (1999). Quality in qualitative research. *Qualitative Inquiry, 5*(4), 465–478. https://doi.org/10.1177/107780049900500402

Semali, L. M., & Kincheloe, J. L. (1999). "Introduction: What is indigenous knowledge and why should we study it." In L. M. Semali & J. L. Kincheloe (Eds.) *What is indigenous knowledge? Voices from the academy*. New York, NY: Routledge.

Serageldin, I. (2013). Ancient Alexandria and the dawn of medical science. *Global Cardiology Science and Practice, 2013*(4), 1–10.

Sharpe, J. (2005). Postcolonial studies in the house of US multiculturalism. In H. Schwartz & S. Ray (Eds.), *A companion to postcolonial studies* (pp. 112–125). Malden, MA: Blackwell Publishing.

Sheets, R. H. (2005). *Diversity pedagogy: Examining the role of culture in the teaching-learning process*. Boston, MA: Allyn & Bacon.

Sheth, M. J. (2018). Grappling with racism as foundational practice of science teaching. *Science Teacher Education, 103*(1), 37–60.

Shizha, E. (2010). The interface of neoliberal globalization, science education and indigenous African knowledges in Africa. *Journal of Alternative Perspectives in the Social Sciences, 2*(1), 27–58.

Singer, J. D., & Braun, H. I. (2018). Testing international assessments. *Science, 360*(6384), 38–40.

Sleeter, C. E. (2001). Preparing teachers for culturally diverse schools: Research and the overwhelming presence of Whiteness. *Journal of Teacher Education, 52*(2), 94–123. https://doi.org/10.1177/0022487101052002002

Smedley, A., & Smedley, B. (2005). Race as biology is fiction, racism as a social problem is real: Anthropological and historical perspectives on the social construction of race. *American Psychologist, 60*, 16.

Smelser, N. J., Wilson, W. J., & Mitchell, F. (2001). *America becoming: Racial trends and their consequences* (Vol. 1). Washington, DC: National Academies Press.

Smith, J., Schellnhuber, H.-J., & Mirza, M. M. Q. (2001). Vulnerability to climate change and reasons for concern: A synthesis. *TAR climate change 2001: Impacts, adaptation, and vulnerability* (pp. 915–967). Cambridge, UK: Cambridge University Press.

Spradley, J. P. (1980). *Participant observation*. New York, NY: Holt, Rinehart and Winston.

Stake, R. E. (1995). *The art of case study research*. Thousand Oaks, CA: Sage Publications.

Stake, R. E. (2005). Qualitative case studies. In N. K. Denzin & Y. S. Lincoln (Eds.), *The Sage handbook of qualitative research* (3rd ed., pp. 443–466). Thousand Oaks, CA: Sage Publications.

Stake, R. E. (2010). *Qualitative research: Studying how things work*. New York, NY: Guilford Press.

Stecchini, L. C. (1961). A history of measures. *The American Behavioral Scientist, 4*(7), 18–21.

Stecchini, L. C. (1971). Notes on the relation of ancient measures to the great pyramid. In T. Tompkin (Ed.), *Secrets of the Great Pyramid*. New York, NY: Harper and Row.

Steinberg, S. R., & Kincheloe, J. L. (2010). Power, emancipation, and complexity: Employing critical theory. *Power and Education, 2*(2), 140. https://doi.org/10.2304/power.2010.2.2.140

Stiglitz, J. E. (2003). *Globalization and its discontents*. New York, NY: W.W. Norton.

Stiglitz, J. (2006). A new agenda for global warming. *The Economists' Voice*, *3*(7). https://doi.org/10.2202/1553-3832.1210

Stone, B. (2020). Understanding the influence of teacher-directed scientific inquiry on students' primal inquiries in two American science classrooms. *Journal of Research in Primary Education*. https://doi.org/ 10.1080/02568543.2020.1718807

St. Pierre, E. A. (2000). The call for intelligibility in postmodern educational research. *Educational Researcher*, *29*(5), 25–28.

Strauss, A. L. (1990). Systematic coding in qualitative research. *Bulletin of Sociological Methodology*, 27(1), 52–62. https://doi.org/10.1177/075910639002700103

Strauss, A. L., & Corbin, J. M. (1990). *Basics of qualitative research: Grounded theory procedures and techniques*. Newbury Park, CA: Sage.

Sunderman, G. L., & Orfield, G. (2006). Domesticating a revolution: No child left behind reforms and state administrative response. *Harvard Educational Review*, *76*(4), 526–556.

Suriel, R. L., & Atwater, M. M. (2012). From the contribution to the action approach: White teachers' experiences influencing the development of multicultural science curricula. *Journal of Research in Science Teaching*, *49*(10), 1271–1295. https://doi.org/10.1002/tea.21057

Swartz, E. (2003). Teaching white preservice teachers: Pedagogy for change. *Urban Education*, *38*(3), 255–278. https://doi.org/10.1177/0042085903038003001

Swartz, E. (2005). Multicultural education: From a compensatory to a scholarly foundation. In C. A. Grant (Ed.), *Research and multicultural education: From the margins to the mainstream* (pp. 31–42). London: Falmer Press.

Tan, E., Barton, A. C., Gutiérrez, M. V., & Turner, E. E. (2012). *Empowering science and mathematics education in urban schools*. Chicago, IL: The University of Chicago Press.

Tate, W. F. (1997). Critical race theory and education: History, theory, and implications. *Review of Research in Education*, 22(1), 195–247. https://doi.org/10.3102/0091732X022001195

Terry, J. (1997). The seductive power of science. In V. A. Rosario (Ed.), *Science and homosexualities* (pp. 271–289). New York, NY: Routledge.

The British Medical Journal. (1927). Imhotep: The physician-architect. *The British Medical Journal*, *1*(3458), 734.

Tigue, K. (2020, April 17). COVID-19 and climate change threats compound in minority communities. *Inside Climate News*. Retrieved from https://insideclimatenews.org/news/16042020/coronavirus-climate-environmental justice oil-refinery-hurricanes-port-arthur-texas

Tolbert, S., Snook, N., Knox, C., & Udoinwang, I. (2016). Promoting youth empowerment and social change in/through social science. *Journal for Activist Science and Technology Education*, 7(1), 52–62.

Toldson, I. & Lewis, C. (2012). *Challenging the status quo: Academic success among school-age African-American males*. Washington, DC: Congressional Black Caucus Foundation, Inc.

Torney-Purta, J., & Amadeo, J. (2013). International large-scale assessments: Challenges reporting and potentials for secondary analysis. *Comparative and International Education, 8*(3), 248–258.

Trends in International Mathematics and Science Study (TIMSS). (2015a). Mathematics for grades 4 and 8: Averages. *National Center for Education Statistics*. Retrieved from https://nces.ed.gov/timss/timss2015/timss2015_table02.asp

Trends in International Mathematics and Science Study (TIMSS). (2015b). Science for grades 4 and 8: Averages. *National Center for Education Statistics*. Retrieved from https://nces.ed.gov/timss/timss2015/timss2015_table23.asp

Tsurusaki, B. K., Barton, A. C., Tan, E., Koch, P., & Contento, I. (2013). Using transformative boundary objects to create critical engagement in science: A case study. *Science Education*, *97*(1), 1–31. https://doi.org/10.1002/sce.21037

Tubbs, N. M. (2005). The philosophy of critical pedagogy. In I. Gur-Ze'ev (Ed.), *Critical theory and critical pedagogy today: Toward a new critical language in education* (pp. 226–240). Haifa, Israel: University of Haifa.

Ültanir, E. (2012). An epistemological glance at the constructivist approach: Constructivist learning in Dewey, Piaget, and Montessori. *International Journal of Instruction, 5*(2), 195–212.

UNESCO. (1974). *Symposium on "The peopling of Ancient Egypt and the deciphering of the Meroitic Script"* (Final Report No. SBC-73/CONF.812/4). Paris, France: United Nations Educational, Scientific and Cultural Organization (UNESCO).

UNESCO. (1978). The peopling of ancient Egypt and the deciphering of the Meroitic script. In: *Proceedings of the symposium held in Cairo from 28 January to 3 February 1974*. Paris, France: United Nations Educational, Scientific and Cultural Organization (UNESCO).

[University of Tesaw] Office of Institutional Research Data. (2014). *Undergraduate degree credit headcount enrollment by major, option, degree, race, and sex for each college: Spring 2014. Table III_6*. University of North Carolina at Charlotte. Retrieved from https://ir.uncc.edu/sites/ir.uncc.edu/files/media/factbook/fb13/fb13036f.pdf

[University of Tesaw] Office of Institutional Research Data. (2017). *Undergraduate degree credit headcount enrollment by major, option, degree, race, and sex for each college: Spring 2017. Table III_6*. University of North Carolina at Charlotte. Retrieved from https://ir.uncc.edu/sites/ir.uncc.edu/files/media/factbook/fb16/fb16036-s.pdf

University Profile. (2019). *Undergraduate Admissions*. University of North Carolina at Charlotte. Retrieved from https://admissions.uncc.edu/about-unc-charlotte/university-profile

U.S. Bureau of Labor Statistics. (2015). STEM crisis or STEM surplus? Yes and yes. *Monthly Labor Review*. Retrieved from https://www.bls.gov/opub/mlr/2015/article/stem-crisis-or-stem-surplus-yes-and-yes.htm

U.S. Census Bureau. (2015). *Projections of the size and composition of the U.S. population: 2014 to 2060: Estimates and projections*. Washington, DC.

U.S. Department of Defense. (2020). *DoD releases fiscal year 2021 budget proposal*. Department of Defense, Washington, DC. Retrieved from https://www.defense.gov/Newsroom/Releases/Release/Article/2079489/dod-releases-fiscal-year-2021-budget-proposal/

U.S. Energy Information Administration. (2019, November 22). Renewable energy explained Incentives—U.S. Energy Information Administration (EIA). *U.S. Energy Information Administration: Independent statistics and analysis*. Retrieved June 12, 2020, from https://www.eia.gov/energyexplained/renewable-sources/incentives.php

Van Dijk, E. M. (2011). Portraying real science in science communication. *Science Education*, *95*(6), 1086–1100. https://doi.org/10.1002/sce.20458

Van Dijk, E. M. (2013). Relevant features of science: Values in conservation biology. *Science & Education*, 22(9), 2141–2156. https://doi.org/10.1007/s11191-012-9496-y

Van Dijk, E. M. (2014). Understanding the heterogeneous nature of science: A comprehensive notion of PCK for scientific literacy. *Science Education*, *98*(3), 397–411. https://doi.org/10.1002/sce.21110

Van Sertima, I. (1984). *The lost sciences of Africa: An overview*. New Brunswick: Transaction Books.

von Glasersfeld, E. (1984). An introduction to radical constructivism. In P. Watzlawick (Ed.), *The Invented reality: How do we know what we believe we know?* (pp. 17–40). New York, NY: Norton.

Wacquant, L. J. D., & Wilson, W. J. (1989). The cost of racial and class exclusion in the inner city. *The Annals of the American Academy of Political and Social Science*, *501*(1), 8–25. https://doi.org/10.1177/0002716289501001001

Wallace, M., Hagan, L., Curran, K. G., Williams, S. P., Handanagic, S., Bjork, A., Davidson, S. L., et al. (2020). COVID-19 in correctional and detention facilities—United States, February–April 2020. *MMWR. Morbidity and Mortality Weekly Report*, *69*(19), 587–590. https://doi.org/10.15585/mmwr.mm6919e1

Wallace, T., & Brand, B. R. (2012). Using critical race theory to analyze science teachers culturally responsive practices. *Cultural Studies of Science Education,* 7(2), 341–374. https://doi.org/10.1007/s11422-012-9380-8

Wallis, C. (2020, June 12). Why racism, not race, is a risk factor for dying of COVID-19. *Scientific American*. Retrieved June 12, 2020, from https://www.scientificamerican.com/article/why-racism-not-race-is-a-risk-factor-for dying-of-covid-19/.

Wallerstein, I. M. (1974). The rise and future demise of the world capitalist system: Concepts for comparative analysis. *Comparative Studies in Society and History*, *16*(4), 387–415.

Wallerstein, I. M. (2004). *World-systems analysis: An introduction*. Durham, NC: Duke University Press.

Wang, W. (2012). The rise of intermarriage: Rates, characteristics vary by race and gender. *Social & Demographic Trends*. Washington, DC: Pew Research Center. Retrieved from http://www.pewsocialtrends.org

Wang, X. (2013). Why students choose STEM majors: Motivation, high school learning, and postsecondary context of support. *American Educational Research Journal*, *50*(5), 1081–1121. https://doi.org/10.3102/0002831213488622

Watson, M. J., & Wiggan, G. (2016). Sankofa healing and restoration: A case study of African American excellence and achievement in an urban school. *Africology: The Journal of Pan African Studies,* 9(1), 59–86.

Weber, M. (2001). *The Protestant ethic and the spirit of capitalism*. New York, NY: Routledge.

Weber, M. (2009). Selected writings. In C. C. Lemert (Ed.), *Social theory: The multicultural and classic readings* (pp. 103–129). Boulder, CO: Westview Press.

Wells, S. (2003). *The journey of man: A genetic odyssey*. New York, NY: Random House Trade Paperbacks.

Wenz, P. S. (1988). *Environmental justice. SUNY series in environmental public policy*. Albany, NY: State University of New York Press.

Whitebook, J. (1979). The problem of nature in Habermas. *Telos*, *1979*(40), 41–69. https://doi.org/10.3817/0679040041

Wiggan, G. (2007). Race, school achievement, and educational inequality: Toward a student-based inquiry perspective. *Review of Educational Research*, 77, 310–333.

Wiggershaus, R. (1995). *The Frankfurt School its history, theories, and political significance*. (M. Robertson, Trans.). Cambridge, MA: MIT Press.

Wilber, K. (2000). *Sex, ecology, spirituality: The spirit of evolution*. Boston, MA: Shambhala.

Wilkinson, R. G. (1996). *Unhealthy Societies: The afflictions of inequality*. London, UK: Routledge.

Wilkinson, R. G. (2007). *The impact of inequality: How to make sick societies healthier*. New York, NY: New Press.

Williams, C. (1987). *The destruction of black civilization: Great issues of a race from 4500 B.C. to 2000 A.D.* Chicago, IL: Third World Press.

Williams, E. E. (1944). *Capitalism & slavery*. Chapel Hill, NC: University of North Carolina Press.

Wilson, E. (1997). Introduction. In T. Abe, S. A. Levin, & M. Higashi (Eds.), *Biodiversity: An ecological perspective* (pp. 1–5). New York, NY: Springer.

Wilson, E. (1998). *Consilience: The unity of knowledge*. New York, NY: Knopf.

Wilson, W. J. (1991). Studying inner-city social dislocations: The challenge of public agenda research: 1990 presidential address. *American Sociological Review*, *56*(1), 1. https://doi.org/10.2307/2095669

Wilson, W. J. (2010). *More than just race: Being black and poor in the inner city*. New York, NY: W. W. Norton.

Winant, H. (2000). Race and race theory. *Annual Review of Sociology*, *26*, 169–185.

Witzig, R. (1996). The medicalization of race: Scientific legitimization of a flawed social construct. *Annals of Internal Medicine*, *125*(8), 675. https://doi.org/10.7326/0003-4819-125-8-199610150-00008

Wolcott, H. F. (1999). *Ethnography: A way of seeing*. Walnut Creek, CA: AltaMira Press.

Woodson, C. G. (1933). *The mis-education of the Negro*. New York, NY: AMS Press.

World Economic Forum. (2020, January). Schools of the future: Defining new models of education for the fourth industrial revolution. *World Economic Forum: Platform for shaping the future of the New Economy and Society*. Retrieved from http://www3.weforum.org/docs/WEF_Schools_of_the_Future_Report_2019.pdf

World Health Organization. (2020, June). WHO Coronavirus Disease (COVID-19) Dashboard. *World Health Organization*. Retrieved June 12, 2020, from https://covid19.who.int

Wüstenhagen, R., Wolsink, M., & Bürer, M. J. (2007). Social acceptance of renewable energy innovation: An introduction to the concept. *Energy Policy*, *35*(5), 2683–2691. https://doi.org/10.1016/j.enpol.2006.12.001

Xie, Y., & Shauman, K. A. (2003). *Women in science: Career processes and outcomes*. Cambridge, MA: Harvard University.

Yam, K. (2020, March 16). Trump tweets about coronavirus using term 'Chinese Virus'. *NBC News*. Retrieved from https://www.nbcnews.com/news/asian-america/trump-tweets-about-coronavirus-using-term-chinese-virus-n1161161

Yancy, C. W. (2020). COVID-19 and African Americans. *JAMA*, *323*(19), 1891. https://doi.org/10.1001/jama.2020.6548

Yin, R. K. (2003). Case study research: Design and methods. *Applied social research methods series* (3rd ed.). Thousand Oaks, CA: Sage.

Yin, R. K. (2012). *Applications of case study research* (3rd ed.). Thousand Oaks, CA: Sage.

Zeichner, K. M. (2003). The adequacies and inadequacies of three current strategies to recruit, prepare, and retain the best teachers for all students. *Teachers College Record*, *105*(3), 490–519. https://doi.org/10.1111/1467-9620.00248

APPENDIX A
GENERAL OBSERVATION PROTOCOL

Observation Protocol

Date:	Time:
Place:	Course ID:
Observation/Code	Reflection

APPENDIX B
INTERVIEW PROTOCOLS

The semi-structured interview protocols in this section were used for preservice teachers and instructors of the courses selected for this study. Questions were tailored to participants accordingly during the interview process.

APPENDIX B.1 PRESERVICE TEACHER INTERVIEW PROTOCOL

SECTION I: Introduction and Warm-up

- **Intro:** *Thank you....*
- Confirm the respondent's name; tell them it won't use again in order to assure anonymity.
- Remind participant: *This study is interested in the role of science in teaching about diversity.*
- ***Begin recording...***
- Give participant a chance to ask questions about the study or the interview process.
- Get verbal consent: *Review informed consent*
- Mention the participants' number which will be randomly generated from emails provided in the questionnaire (i.e., you are participant #24245).

Warm-up Questions

- Tell me a little about where are you from. (*What it's like; people you grew up with, etc.*).
- What's your major?
- How far along are you?

SECTION II: Race, Culture, and Education

1 How would you describe your cultural background?
2 What has been most influential in shaping your views of human diversity, and more generally, people who are different from you?
3 Can you provide an example of a classroom experience where you were discussing racial/ethnic issues with a person you perceived to be different than you? (Good or bad).

   - Can you provide a specific example?

4 Describe your thoughts about teaching students who may be culturally different than you.

   - That is, how will student diversity be addressed in your classroom?
   - How confident are you in your ability to teach racially, ethnically, and culturally diverse students?

5 In your view, what effect does a child's cultural background have on his or her education?

   - What has informed your response?

6 Describe what is needed to be a successful teacher of students from diverse backgrounds... (*How did you arrive at this conclusion?*)

SECTION III: Science, Diversity, and Education

1 In what ways has science or scientific knowledge been influential in your life and understanding the world (e.g., how you live, work, play, learning, etc.)?
2 What can science or scientific knowledge tell us about race, ethnicity, and culture (if anything)?

   - What can science contribute to understanding race relations and multiculturalism in society?
   - What are your thoughts on science contributing to inclusive, pluralistic schools and society?

3 How might the use of science help PSTs better understand issues of racial and ethnic diversity?

   - In what ways could doing so inform your practice in positive and inclusive ways?
   - How might you use scientific knowledge to start a discussion about race and ethnicity?

4 How might using science assist in preparing preservice teachers to address and/or take a stand for human rights and social justice issues?
5 How and/or why could science be used to teach about diversity in teacher education programs (regardless of discipline)?

   - Do feel like you need to be a science teacher to bring this into the classroom or share it with fellow teachers?

Closing

Thank you for agreeing to participate in this study.

Before we conclude, is there anything that you feel was not addressed in this interview that should have been, or anything you would like to add?

Inform participants that they will be offered the chance to review the interview transcripts after they are completed.

Again, thank you for participating in this study

APPENDIX B.2 PROFESSOR INTERVIEW PROTOCOL

SECTION I: Introduction and Warm-up

- **Intro**: *Thank you…*
- Confirm the respondent's name; tell them it won't use again in order to assure anonymity.
- Remind participant: *This study is interested in the role of science in teaching about diversity.*
- **Begin recording…**
- Give participant a chance to ask questions about the study or the interview process.
- Get verbal consent: *Review informed consent*
- Mention the participants' number which will be randomly generated from emails provided in the questionnaire (i.e., you are participant #24245).

Warm-up Questions

- So tell me a little about where are you from. (*What it's like; people you grew up with, etc.*).
- What's your major?
- How far along are you?

SECTION II: Race, Culture, and Education

1 How would you describe your cultural background?
2 What has been most influential in shaping your views of human diversity, and more generally, people who are different from you?
3 Can you provide an example of a classroom experience where you were discussing racial/ethnic issues with a person you perceived to be different than you? (Good or bad).

   - Can you provide a specific example?

4 Describe your thoughts about teaching students who may be culturally different than you.

   - That is, how will student diversity be addressed in your classroom?
   - How confident are you in your ability to teach racially, ethnically, and culturally diverse students?

5 In your view, what effect does a child's cultural background have on his or her education?

   - What has informed your response?

6 Describe what is needed to be a successful teacher of students from diverse backgrounds… (*How did you arrive at this conclusion?*)

SECTION III: Science, Diversity, and Education

1 In what ways has science or scientific knowledge been influential in your life and understanding the world (e.g., how you live, work, play, learning, etc.)?
2 What can science or scientific knowledge tell us about race, ethnicity, and culture (if anything)?

   - What can science contribute to understanding race relations and multiculturalism in society?
   - What are your thoughts on science contributing to inclusive, pluralistic schools and society?

3 How might the use of science help PSTs better understand issues of racial and ethnic diversity?

   - In what ways could doing so inform your practice in positive and inclusive ways?
   - How might you use scientific knowledge to start a discussion about race and ethnicity?

4 How might using science assist in preparing preservice teachers to address and/or take a stand for human rights and social justice issues?
5 How and/or why could science be used to teach about diversity in teacher education programs (regardless of discipline)?

- Do feel like you need to be a science teacher to bring this into the classroom or share it with fellow teachers?

Closing

Thank you for agreeing to participate in this study.

Before we conclude, is there anything that you feel was not addressed in this interview that should have been, or anything you would like to add?

Inform participants that they will be offered the chance to review the interview transcripts after they are completed.

Again, thank you for participating in this study.

APPENDIX C
LESSON PLAN

Science and Human Diversity:
The Empirical Challenges of Racial Classification
Adapted from a lesson developed by Scott Bronson
[*Used for CPSP references and resources*]

Subject Matter: Biology, Anthropology, Genetics, Geography
Time Allotment: 1–2 class sessions
Description: This lesson will help students examine their preconceptions and assumptions about racial categories and understand the impossibility of constructing a consistent system of human racial classification.

Overview

Most people still believe that the world's people come divided into separate, biologically distinct groups called "races," distinguished by physical characteristics such as skin color, eye shape, and hair texture. The idea of race assumes that variation in these superficial traits correlates with other innate differences, including intelligence and aptitude – in other words, that members of one race are fundamentally more similar to each other than to members of another race.

Human beings across the world do in fact vary in their physical appearance and genetic composition. And traits like skin color, eye shape and hair texture are influenced by genes we inherit from our parents. But do patterns of human variation map onto "races"? Why do we classify the way we do, using some traits while ignoring others? Is there a logically consistent way of assigning individuals to a biological "race"?

This lesson will help students discover the fuzzy, inconsistent nature of racial classification – including how and why patterns of human variation defy racial categorization – and learn about some of the hidden values and assumptions that underlie the reasons we classify the way we do.

Students will first explore their own preconceptions about race by sorting themselves according to which race they think they belong to and listing the criteria they used. Students will then be asked to re-sort themselves according to several of their own observed genetic traits. These results will be compared within and between "races" to explore the impossibility of consistent human racial classification.

Glossary

gradual variation	non-concordance
within-group vs. between group variation	genotype
phenotype	natural selection
sexual selection	gene flow
polymorphism	Allele

Objectives

1. Confront the inconsistencies inherent in any attempt to classify humans according to innate racial categories.
2. Better understand patterns of human variation, including how the geographic distribution of physical traits does not follow racial lines but rather cuts across them.
3. Introduce three basic concepts of human biodiversity: gradual variation, non-concordance and within-group vs. between-group variation.
4. Understand the importance of logical consistency in scientific reasoning and experiments.
5. Contribute information to class discussions.

Materials

General

RACE – The Power of an Illusion, Episode 1 – The Difference Between Us

- RACE companion Web site www.pbs.org/race:
  - Sorting People interactivity + Go Deeper
  - Human Diversity interactivity + Go Deeper
- Human Traits Worksheet; Sickle Cell and Skin Color Handout (both attached)
- DNAI Web site

Optional Readings

Jonathan Marks, "Scientific and Folk Ideas about Heredity": http://www.uncc.edu/jmarks/interests/Baltimore.html

Jared Diamond, "Races without Color," Discover Magazine, 11/94 (www.discover.com/archive)

Stephen Jay Gould, "The Geometer of Race," Discover Magazine, 11/94 (www.discover.com/archive)

Lesson Plan Activities

Activity #1

Articulating Preconceptions: What Do Students Believe?

Brainstorm the different races of the world and list all the "races" across the whiteboard. Students might use different and inconsistent kinds of labels for their races (see below). This should not be discouraged.

If possible, have students physically group themselves in the classroom according to the race they think they belong to. When everyone has joined a group, ask each group to generate a list of the criteria for membership in their "race" (e.g., skin color), and post the list on the whiteboard under their respective racial label.

NOTE: Some students may resist joining any of the designated racial groups. That's okay, but they must then add whatever label each would apply to him or herself to the list of racial labels, even "human being."

As a class, briefly note and discuss any inconsistencies across racial labels – for example, categories based on anthropology (Mongoloid, Caucasian, Negroid) vs. color (Yellow, White, Brown, Black) vs. geography (Asian, European, African) vs. ethnicity (Arab, Jew, Kurd, Hmong) vs. nationality (Mexican, Puerto Rican) vs. language (Hispanic), etc. – and similarly discuss the membership criteria listed on the board. Are they all comparable?

Alternate Activity #1

For Racially Homogenous Classes

Ask your students to do the "Sorting People" activity, below.

This interactivity challenges users to "sort" 20 people (images provided) into five federally recognized categories. Once they finish sorting, students learn (a) how the U.S. government would classify these people; (b) how each person self-identifies; and (c) each person's ancestry. Students can then move on to the "Explore Traits" activity where the same 20 people are re-sorted according to skin color, blood type and fingerprint type.

Activity #2

Problematizing Student Assumptions: Creating an Inventory of Traits

Screen the first five minutes of Episode 1 – *The Difference Between Us*, beginning with the opening titles (00:45–05:23; DVD Scene #2) – this is **Clip A.**

Stop the video and ask students if they too believe their closest genetic matches are with others in their own racial group. As a class activity, ask your students to brainstorm definitions of "race." List each definition on the board.

If possible, pass out the Human Traits Inventory Worksheet (below), which lists several traits influenced by our genes. Have the students work in small groups to analyze their own physical traits and fill out the worksheet. Otherwise, discuss this with the class as a whole by show of hands.

Human Traits Inventory Worksheet

Inherited Trait	*Yes #*	*No #*
Hitchhiker's thumb (top joint can form almost 90-degree angle)		
Tongue curl		
Detached earlobe		
Skin color – Dark		
Skin color – Medium		
Skin color – Light		
Widows peak		
Hand clasp – Right thumb on top		
Hand clasp – Left thumb on top		
Fingerprint type (click here for examples) – Loop		
Fingerprint type – Whorl		
Fingerprint type – Arch		
Fingerprint type – Tented Arch		
Blood type A (ask parent, doctor, or clinic)		
Blood type B		
Blood type AB		
Blood type O		
Lactose intolerant		

If previously grouped, have students re-form their original "racial" groups. Then, proceeding down the list of traits on the worksheet, ask the students to physically re-group themselves in the classroom by each successive trait (those with hitchhiker thumb on one side of the room, those without on the other side, etc.).

After going through the entire list, discuss the following: What do the students notice about the new groupings? Do any of them match up or co-vary with their original racial groupings? How would students divide themselves now on the basis of genetic similarity and difference, taking into account all of these traits? Would it be easier to categorize people according to race if we used more traits? Why or why not?

Now, have the class re-visit the original list of membership traits for each "racial" group generated earlier (these should already be listed on the whiteboard). Discuss the following: Why did they choose some traits but ignore others when dividing into racial groups? Why do we assume that traits like skin color are more meaningful than whether or not your tongue curls or whether or not your earlobes are attached? What do students think of their earlier definitions of race?

NOTE: Many students may be troubled and confused as they begin to scrutinize the "commonsense" racial categories we have all taken for granted but never examined. This cognitive dissonance is to be expected. You might re-assure students that we're not claiming that race isn't "real" (see discussions of "social race" at the end of Episode 1 and the other two episodes), but that the biological criteria we use are fuzzy, inconsistent and don't correspond to real patterns of human variation.

At this point, if it is a diverse class, perhaps re-visit the common racial skin color labels (red, yellow, white, brown). Have the class divide into groups of eight or so, and have them compare their skin colors using the inside of their upper arms (as is done in the film). What do they find? Are all "Black" people really darker than all "White" people? What about individuals from East Asian, Latino and Native American backgrounds? South Asians? What color are they really? (This can be done even with a racially homogenous class, but is best paired with the "Explore Other Traits" activity in the Sorting People activity).

Activity #3

Mapping Traits onto Racial Categories

Summarize Jared Diamond's "Race without Color," from Discover Magazine (www.discover.com/archive), which includes a discussion of how various inherited traits are distributed geographically.

If possible, have students work in small groups to make a chart that groups populations into "races" for each of the following traits. Otherwise, use whole class discussion format and show of hands (NOTE: advanced classes may also want to research these traits in other populations for a more complete picture):

- Dark skin/medium skin/light skin
- Lactose tolerance/intolerance

- Epicanthal eye-folds/no epicanthal eye-folds
- Sickle-cell carriers/not sickle cell carriers
- Shovel-shaped incisors/other incisors

Include here a discussion about Pangaea and the influence of plate tectonics (continental drift) on the current geography of the continents. This will help put into context later discussions about phenotypic differences among humans. Then, using different colored pins and a photocopy of a world map mark the geographical distribution of one or more of the traits.

As a prelude to discussing the class's findings, ask students to define "genotype" and "phenotype." Does phenotype always reflect genotype? What other factors might affect phenotype besides the genes? Note that with the exception of sun-tanning on skin color, the phenotypic traits above do closely reflect genotype. Emphasize that very few traits are influenced by only one gene. Most traits are influenced by many genes, interacting with each other and with the environment in a very complex dance that scientists have barely begun to understand.

Discuss the class's findings. What surprised students the most? Does the geographic distribution of these traits match up and co-vary with what we conventionally think of as races? Or do the traits cut across racial lines?

For further discussion, reference the Human Diversity section (select "Physical Appearance") of the RACE Web site to see how some of the traits we usually think of as racial are distributed geographically.

Activity #4

Why Racial Classification Doesn't Work

Understanding Gradual Variation, Non-Concordance and Within- vs. Between-Group Variation

We've now seen how difficult it is to find a consistent racial classification system that works for everyone. Show the following two video clips. As they watch, ask students to look for three underlying characteristics of human variation that explain why racial classification doesn't work.

CLIP B: (5 minutes) – on skin color and gradual variation; which begins with Alan Goodman on the difficulty of measuring race, ends with Joseph Graves taking us on an imaginary walk from the tropics to the pole (22:35–28:03; DVD Scene #8 [back up 20 seconds from the start of Scene #8 to include the statement by Alan Goodman about the difficulty of measuring race])

CLIP C: (9 minutes) – on non-concordance and within-group vs. between-group variation. Begins with Alan Goodman talking about concordance, ends with the narrator explaining that sickle cell results from having ancestors who lived in malarial regions. (32:20–40:46; DVD Scenes #10, #11, #12)

Before moving on to a discussion of why racial classification doesn't work, make certain your students understand sickle cell trait and the skin color stories told in the film using the following handouts, as necessary.

NOTE: For an excellent lab on how selective forces affect allele frequencies using the example of sickle cell, visit the Genetics Project Web site: http://chroma.mbt.washington.edu/outreach/genetics/sickle/sicklebean.html

Handouts

Sickle Cell Handout

Sickle cell anemia is a hereditary disease that arises from a single mutation (SNP or single nucleotide polymorphism) in one of the genes that code for the hemoglobin protein in our red blood cells. Hemoglobin carries oxygen through the bloodstream to the body.

People who inherit two copies of the mutated hemoglobin gene, one from each parent, get sickle cell anemia. Their red blood cells become sticky and stiff and sometimes become sickle shaped. These "sickled" cells tend to get stuck in the narrow blood vessels known as capillaries, blocking the flow of blood. Sickle cell anemia is a very painful disease and can be life threatening.

But a larger number of people who inherit just a single copy of the mutated gene are known as sickle-cell carriers. They usually don't become anemic and interestingly, sickle cell carriers tend to be resistant to malaria, a deadly disease.

Malaria is caused by a parasite carried by the Anopheles mosquito. After humans started practicing agriculture, the mosquito thrived in standing pools of water that appeared on the cleared land. Scientists believe that the sickle cell mutation arose independently four or five different times in human history, no more than 10,000 years ago, and probably much more recently. Because carrying one sickle cell gene conferred a survival advantage in areas of the world where malaria was common, the sickle cell mutation was positively selected and passed on in those regions.

As humans migrated, the sickle cell trait spread, not by contagion, of course, but through reproduction. Population geneticists call it "gene flow" when a gene variant like sickle cell passes from one population to another. As a result of these historical and environmental influences, sickle cell is commonly found in central and western Africa, but not Southern Africa. It is also found on the Arabian Peninsula and over into India, as well as up through Turkey, Greece, Albania, Sicily, and Italy and other parts of the Mediterranean basin.

Carrying the sickle cell variant of the hemoglobin gene, therefore, is a marker not of race, but of descent from people who once lived where malaria was common.

Show map of the spread of sickle cell.

NOTE: another mutation on the same gene but at a different locus also confers resistance to malaria, and in its double form also causes a blood disease. This disease is known as thalassemia. Thalassemia is most common in parts of southern/southeastern Asia and southern Europe.

Skin Color Handout

Why do we have different skin colors? Scientists can only hypothesize at this point because the genetics of skin color is poorly understood. But we do know that skin color tends to correlate not with "race" but with the amount of ultraviolet radiation in the environment.

People in equatorial areas of the world, where sunlight is most intense, have the darkest skin. This includes sub-Saharan African peoples; Tamils, Dravidians, and others from southern India and Sri Lanka; Aborigines in Australia; and Melanesians in the South Pacific. People living in areas where sunlight is less intense, further north and south, tend to have lighter-colored skin.

Scientists don't know why this is, but they have several theories. One theory highlights the role of vitamins and natural selection. The reasoning is as follows:

Dark melanin (skin color variation is controlled by the kind and amounts of pigment in our skin called melanin) acts as a natural sunscreen, blocking ultra-violet (UV) radiation. Too much UV radiation destroys an important enzyme made under the skin called folate. If pregnant mothers don't get enough folate, their infants may be born without a full brain, spinal cord or with other neural-tube disorders. By this reasoning, anthropologist Nina Jablonski suggests that dark skin would be a selective advantage in the equatorial areas of the world because it would block more UV radiation and thus prevent the destruction of folate (see "A New Light on Skin Color" in Resources, below). Thus, darker-skinned people would be better able to survive and reproduce in these areas.

On the other hand, the body needs some UV radiation to manufacture Vitamin D under the skin. Without enough Vitamin D, humans are prone to rickets, a crippling bone disorder most common in young children. Scientists hypothesize that among groups of early humans who migrated to regions far from the equator, where there was less sunlight, those individuals with lighter skin were better able to survive and reproduce (because they could absorb enough UV radiation to produce adequate amounts of Vitamin D). Thus light skin was increasingly selected for in those areas. (Today, most of us get Vitamin D from fish oil, cereals, and fortified milk – foods not available to early humans until we developed fishing and farming technologies).

Over time, these two contrasting influences created a marked difference in the appearance of people living in areas with different amounts of UV radiation.

NOTE: Some people assume that dark melanin provides a selective advantage because it protects against skin cancer by blocking UV radiation. While dark skin does protect against skin cancer, it does NOT provide a selective advantage. That's because skin cancer doesn't usually manifest until later in life, after the childbearing years. Unless a trait affects the ability to reproduce, no natural selection will occur and traits – positive or negative – will continue to be passed down.

An alternative theory (favored by Darwin) suggests that sexual selection drove the evolution of different skin colors and the distribution of other surface traits like hair form. According to this scenario, certain cultures (for one reason or another) came to value dark skin while others valued lighter skin color pigments. Individuals with the favored appearance in a particular culture would have the most opportunities to mate and have more children, just as peacocks with the biggest tail feathers are favored by peahens and thus out-reproduce other peacocks. Over time, this skewed reproduction would result in differences between cultures that preferred different traits. The advantage to this theory is that sexual selection can spread a favored trait through a population very quickly, while natural selection works more slowly.

Now, based on the film clips, ask the class to identify and describe three characteristics of human variation that would explain why it is so difficult to classify humans into racial boxes.

The three characteristics are:

- ***Gradual Variation.*** Variation of traits is generally continuous, with no clear boundaries between them. As Joseph Graves illustrates in Episode 1 (through his imaginary walk), it is impossible to say where one race ends and another begins. Because of gene flow, groups living close to each other tend to be more visibly similar, while groups far away tend to be less so. Geography is a better explanation for similarity than race.
- ***Non-Concordance.*** Human variation is highly non-concordant. One trait rarely co-varies with or predicts for another. This is because most traits are influenced by different genes, and genes are inherited independently one from another. Knowing a person's skin color does not help you predict other traits, such as height, blood type, or sickle cell. As Alan Goodman says, "Skin color or eye color or hair color is not correlated with height or weight. And they're definitely not correlated with more complex traits like intelligence

or athletic performance." In a nutshell, racial profiling doesn't work on a genetic level.

- ***Within-Group versus Between-Group Variation.*** Most of the world's diversity can be found in any local population. As Richard Lewontin first observed, on average, 85% of all human variants can be found within any local population, be they Swedes, Hmong, or Fulani. (About 93–96% can be found on any continent, which means only about 4–7% of human variation is explained by continental divisions). In other words, as Lewontin points out in his book Human Diversity, if a neutron bomb tomorrow wiped out everyone in the world except for Nelson Mandela's Xhosa people, 85% of the world's genetic variation would still be left, though the remaining population would on average be darker skinned. Because of this huge within-group diversity, knowing an individual's purported race tells us little about his or her specific genes and traits.

Activity #5

Reading and Class Discussion

As a final exercise, summarize and discuss the following article about our myths and misconceptions about racial classification and its inherent subjectivity.

Jonathan Marks, "Scientific and Folk Ideas about Heredity."

Students may also want to read the "Go Deeper" article in the Sorting People section of the RACE web site.

As a culminating activity, ask students to write an essay addressing the following questions (this essay will be used for assessment):

Consider this conventional definition of race: "The idea of race assumes humans come divided into several distinct groups, each of which carry a set of innate traits. Those who are members of one race are more genetically similar to each other, and more different from members of another race."

Is this definition scientifically sound given the results from the student activity, the film clips and the readings? Why or why not? What basic scientific criteria have to be met for this definition to work? Have your ideas about race been changed by this exercise? Explain.

Assessment

Any assessment will be at the professors' discretion. Students may be evaluated based on their participation in class exercises and the thoroughness of analyses during class discussions. Film clips and topics used during class discussion as well as how students used these resources to support their positions can be used as a guide for the instructor's assessment.

Extensions

Planned for Day #2 (if time) view clips and discuss Spencer Wells documentary "The Journey of Man: A Genetic Odyssey" at: http://youtu.be/OV6A8oGtPc4

Summary

> "By analyzing DNA from people in all regions of the world, geneticist Spencer Wells has concluded that all humans alive today are descended from a single man who lived in Africa around 60,000 years ago"

National Geographic: http://news.nationalgeographic.com/news/2002/12/1212_021213_journeyofman.html

Read/Discuss "From Africa to Astoria by Way of Everywhere" on National Geographic website, which discusses Wells' work in relation to Astoria Queens, New York – one of the most ethnically diverse communities on Earth.

Discussions will take place in the context of racial, ethnic, and cultural differences in the classroom and what science is able to tell us about human diversity.

Read Stephen Jay Gould, "The Geometer of Race." Discover Magazine (November, 1994). Gould tells the fascinating story of how White people came to be named after a mountain range in Georgia, and how German naturalist Friedrich Blumenbach's visual model of five races became popularized and inadvertently fueled notions of racial superiority and inferiority.

Students can explore human genetic diversity further using the Cold Spring Harbor Laboratories computer exercise titled "How Unique Are You?" This exercise will calculate how many other students, out of thousands that have participated in this exercise, have the same genetic profile that they do. Don't be surprised if the answer is zero. Go to the DNAI Web site (http://www.dnai.org/d/index.html) and navigate to the activity by clicking on Human Origins, then Variation, then Interactive Variation Activity.

Explore or discuss in class the related essay taken from the "Go Deeper" article associated with the "Sorting People" section of the RACE Web site. Also read the Background Reading on the history of the Census:

Should We Classify?

Following are the U.S. federal government's current definitions for the racial and ethnic groups we used in the sorting activity:

- ***American Indian or Alaskan Native.*** A person having origins in any of the original peoples of North and South America (including Central America), and who maintains tribal affiliation or community recognition.

- ***Asian.*** A person having origins in any of the original peoples of the Far East, Southeast Asia, the Indian subcontinent including, for example, Cambodia, China, India, Japan, Korea, Malaysia, Pakistan, the Philippine Islands, Thailand, and Vietnam.
- ***Black or African American.*** A person having origins in any of the Black racial groups of Africa. Terms such as "Haitian" or "Negro" can be used in addition to "Black or African American."
- ***Hispanic or Latino.*** A person of Cuban, Mexican, Puerto Rican, South or Central American, or other Spanish culture of origin, regardless of race. The term "Spanish origin" can be used in addition to "Hispanic or Latino."
- ***Native Hawaiian or Other Pacific Islander.*** A person having origins in any of the original peoples of Hawaii, Guam, Samoa, or other Pacific Islands.
- ***White.*** A person having origins in any of the original peoples of Europe, the Middle East, or North Africa.

Most of these categories were introduced in 1977, in response to new civil rights laws designed to remedy discrimination. Look closely at these definitions. Is everybody defined in the same way? To be categorized as Native American, for example, requires "tribal affiliation or community recognition" – a condition of no other category. The definitions for African American includes a reference to "Black racial groups," while none of the other categories mention race. In fact, Hispanic or Latino is defined as a "Spanish culture of origin, regardless of race." The category Native Hawaiian or Other Pacific Islander was only introduced in 1996 – previously, it was lumped together with Asians.

What reasons might exist for defining these groups in these seemingly contradictory ways? Are the criteria social or scientific?

Even though there's no consistently objective way to classify people, can you think of reasons why we would want the government to categorize or track information on racial groups? What would happen to efforts to remedy discrimination and inequality if we didn't have any racial data?

Additional Resources

RACE Web Site

- Sorting People
- Sorting People Go Deeper
- Human Diversity
- Richard Lewontin, Human Diversity, Scientific American Library, 1982, 1995

- Stephen Jay Gould, The Mismeasure of Man, W.W. Norton, 1981, 1996
- Saadia Iqbal, "A New Light On Skin Color," National Geographic Magazine Online (http://magma.nationalgeographic.com/ngm/0211/feature2/online_extra.html)
- DNA From the Beginning (http://www.dnaftb.org/dnaftb/)
- DNAi Web site: (http://www.dnai.org/d/index.html)

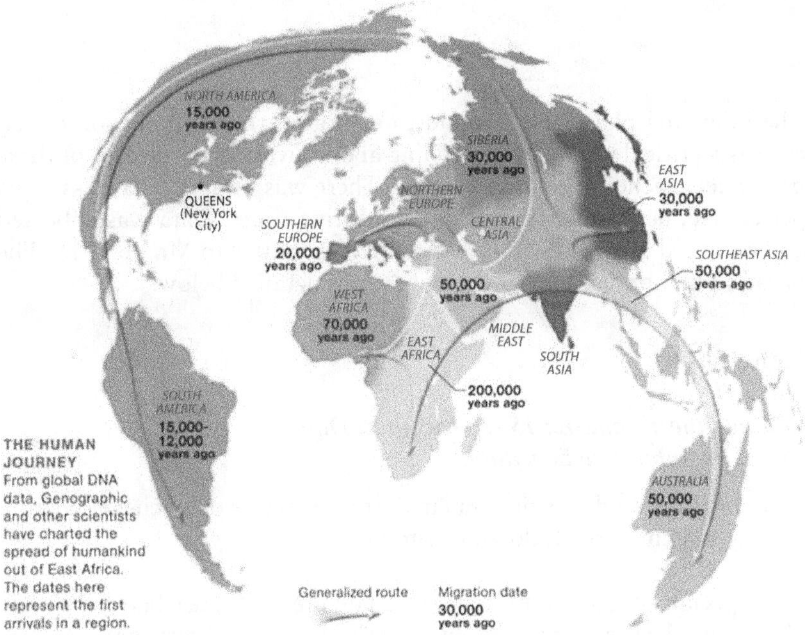

Figure A.1 Human migration map

Source: http://ngm.nationalgeographic.com/img/big-idea/queens-genes.jpg

APPENDIX D
PHASES OF THE STUDY

The following phases correspond to exploring the phenomenon of science as a critical method for teaching about diversity in the case of three preservice teacher education courses. There was total of 3–5 class visits per course, depending on scheduled meeting times. Data was collected and analyzed during the months of November 2014 to March 2015. The duration and procedures in each phase are detailed below.

Phase 1

Online Questionnaire: 15–20 Minutes; Dates/ Times for Interview Scheduled

Prerequisites: Send recruitment emails to professors of teacher education courses matching the following criteria:

a especially designed for undergraduate preservice teachers;
b required either for admission into the TEP or for licensure;
c emphasize social issues in education, i.e., student diversity and teaching in diverse settings.

The voluntary questionnaire was administered online to students and professors in each of the three courses, and included a brief description of the study and an informed consent agreement. Dates/times for the interview were also determined using the form. Some observations and preliminary data analysis began during this phase.

The primary goal of Phase 1 was to collect and begin analyzing participants' demographic information and initial views on human differences, with respect to science, diversity (i.e., race, ethnicity, culture, etc.), and teaching to diverse student populations. This data informed the observations, class discussions and interviews, and was later used for triangulation to help ensure the accuracy and trustworthiness of the study.

Phase 2

Observations: 2 × 75 Minutes for Biweekly Classes; 1 × 150 Minutes for Weekly Classes

Classroom observations began during Phase 2. Observation field notes included rich descriptions of each setting, participant responses and behaviors, in order to provide context to the observation experience. The field notes also used codes/symbols to represent behaviors, etc. This provided a thorough record using fewer words, which allowed more data collection for later analyses (LeCompte, 2000; Spradley, 1980).

Phase 3

Researcher-Led Discussion and Participant Observation: 1 × 75 Minutes

At each instructor's discretion at least one researcher-led discussion was conducted with the entire class. The discussion consisted of various instructional methods, including scientific information that helped explain elements of human diversity (Appendix C). As the lesson(s) aligned with course content, they did not impinge on class time or on students who chose not to participate in other components of the study (i.e., questionnaire or interview). Participant observations and data analysis continued during this time.

Phase 4

Semi-Structured Follow-up Interview: Approx. 45–60 Minutes

In this phase, semi-structured interviews were conducted with 14 of the 30 students who completed the questionnaire as well as two course instructors. The interviews were held at the convenience of participants either by phone, Skype, or in person, and were guided by protocols in Appendix B. The semi-structured format allowed for a variety of responses, which were audio-recorded, transcribed, and then coded and analyzed for emergent themes. Concluding analyses were conducted throughout and following Phase 4 (Spring 2015). The final two chapters were written during this time.

APPENDIX E
SAMPLING FRAME COURSE DESCRIPTIONS

Education

EDUC 2100. Introduction to Education and Diversity in Schools. Social, historical, and philosophical foundations of the educating professions, the organization and various levels of education, and the major issues in American education. Field-based activities in observing in-class and non-classroom settings: 5 hours.

Middle, Secondary, and K-12 Education

MDSK 2100. Diversity and Inclusion in Secondary Schools. Introduction to the contexts, challenges, and changes in U.S. secondary education; emphasis on diversity in student populations including special needs students; examination of diversity in schools, including school organizations, approaches to teaching, purposes and expectations for public education, and communities. 10 hours of observations and participation in school settings required.

Index

www.ingramcontent.com/pod-product-compliance
Lightning Source LLC
LaVergne TN
LVHW021125300926
849160LV00010B/862
* 9 7 8 0 3 6 7 6 2 1 5 8 2 *